Biomath
Problem Solving
for Biology Students

Robert W. Keck

Indiana University, Purdue University at Indianapolis, Department of Biology

Richard R. Patterson

Indiana University, Purdue University at Indianapolis, Department of Mathematical Science

An imprint of Addison Wesley Longman, Inc.

San Francisco · Reading, Massachusetts · New York · Harlow, England
Don Mills, Ontario · Sydney · Mexico City · Madrid · Amsterdam

Executive Editor: Erin Mulligan
Sponsoring Editor: Elizabeth Fogarty
Project Editor: Erika Buck
Publishing Assistant: Chriscelle Merquillo
Production Editor: Vivian McDougal
Copy Editor: Jan McDearmon
Proofreader: Martha Ghent
Text Designer: Kathleen Cunningham Design
Illustrator: Shirley Bortoli
Compositor: Peter Vacek, Eigentype Compositors
Cover Designer: Yvo Riezebos Design

ISBN 0-8053-6524-9

2 3 4 5 6 7 8 9 10—MV—03 02 01 00

Addison Wesley Longman, Inc.
1301 Sansome Street
San Francisco, CA 94111

Of Related Interest from the Benjamin/Cummings Series in the Life Sciences

General Biology

N. A. Campbell, J. B. Reece, and L. G. Mitchell
Biology, Fifth Edition (1999)

N. A. Campbell, L. G. Mitchell, and J. B. Reece
Biology: Concepts and Connections, Third Edition (2000)

R. A. Desharnais, J. R. Bell, and M. Palladino
Biology Labs On-Line (2000)

J. Dickey *Laboratory Investigations for Biology* (1995)

J. B. Hagen, D. Allchin, and F. Singer
Doing Biology (1996)

A. E. Lawson and B. D. Smith
Studying for Biology (1996)

J. Lee *The Scientific Endeavor* (2000)

J. G. Morgan and M. E. Carter
Investigating Biology, Third Edition (1999)

J. A. Pechenik
A Short Guide to Writing about Biology, Third Edition (1997)

G. I. Sackheim
An Introduction to Chemistry for Biology Students, Sixth Edition (1999)

R. M. Thornton *The Chemistry of Life* CD-ROM (1998)

Biotechnology

D. Bourgaize, T. Jewell, and R. Buiser
Biotechnology: Demystifying the Concepts (2000)

Cell Biology

W. M. Becker, L. J. Kleinsmith, and J. Hardin
The World of the Cell, Fourth Edition (2000)

Biochemistry

R. F. Boyer
Modern Experimental Biochemistry, Third Edition (2000)

C. K. Mathews, K. E. van Holde, and K. G. Ahern
Biochemistry, Third Edition (2000)

Genetics

R. J. Brooker
Genetics: Analysis and Principles (1999)

J. P. Chinnici and D. J. Matthes
Genetics: Practice Problems and Solutions (1999)

R. P. Nickerson
Genetics: A Guide to Basic Concepts and Problem Solving (1990)

P. J. Russell
Fundamentals of Genetics, Second Edition (2000)

P. J. Russell
Genetics, Fifth Edition (1998)

Molecular Biology

M. V. Bloom, G. A. Freyer, and D. A. Micklos
Laboratory DNA Science (1996)

J. D. Watson, N. H. Hopkins, J. W. Roberts, J. A. Steitz, and A. M. Weiner
Molecular Biology of the Gene, Fourth Edition (1987)

Microbiology

G. J. Tortora, B. R. Funke, and C. L. Case
Microbiology: An Introduction, Sixth Edition (1998)

Anatomy and Physiology

E. N. Marieb
Essentials of Human Anatomy and Physiology, Sixth Edition (2000)

E. N. Marieb
Human Anatomy and Physiology, Fourth Edition (1998)

E. N. Marieb and J. Mallatt
Human Anatomy, Second Edition (1997)

Ecology and Evolution

C. J. Krebs
Ecological Methodology, Second Edition (1999)

C. J. Krebs
Ecology: The Experimental Analysis of Distribution and Abundance, Fourth Edition (1994)

J. W. Nybakken
Marine Biology: An Ecological Approach, Fourth Edition (1997)

E. R. Pianka
Evolutionary Ecology, Sixth Edition (2000)

D. A. Ross
Introduction to Oceanography (1995)

R. L. Smith
Ecology and Field Biology, Fifth Edition (1996)

R. L. Smith and T. M. Smith
Elements of Ecology, Fourth Edition Update (2000)

Plant Ecology

M. G. Barbour, J. H. Burk, W. D. Pitts, F. S. Gilliam, and M. W. Schwartz
Terrestrial Plant Ecology, Third Edition (1999)

Zoology

C. L. Harris *Concepts in Zoology*, Second Edition (1996)

Preface

To the Student

Are you curious about

- how much head start a gazelle needs to outrun a cheetah?

- which has the most surface area: your skin, your lungs, or your kidneys?

- how much brain food you get from a candy bar?

These and many other quantitative questions are posed in this book. Working through them will help you appreciate the usefulness of quantitative calculations in biology. You may often be surprised by the answers you find.

Biologists need to know a certain amount of mathematics. Sometimes a calculation is needed to test alternative hypotheses. Numbers and measurements are necessary to make a convincing argument about a biological fact. You will be more confident of results you have computed rather than just accepted as dogma. And finally, you will make yourself more marketable by being able to do computations.

The mathematical skills required to do these problems are mostly at the high school level: conversions of units; algebra; geometry of circles, spheres, and cylinders; logarithms; and graphing. In a few cases, more advanced material is presented in the problem. A Math Review section is provided in Appendix A for you to review the necessary mathematics. Appendix B supplies unit conversion factors.

The biology facts needed to do the problems are largely covered in the problems, but feel free to consult any reference text to fill in your knowledge.

Answers to all problems are included. But *beware!* When we provided answers, we had to make certain assumptions. Other assumptions you might make may be perfectly logical and will likely lead to different numbers. So just because your answer is different does not necessarily mean it is wrong. It may mean you should compare different assumptions.

To the Instructor

The problems in this book are classified by biological topic. In each section, some problems are harder than others. But the mathematics used is chosen for the problem and does not get progressively more difficult. Thus, the topics can be covered in any order.

The problems were designed to achieve several goals:

- demonstrate the contribution of mathematics in quantitative biological reasoning

- develop skills in reading, questioning, analyzing, and evaluating assumptions (critical thinking)

- develop written presentation skills

- foster group interaction.

These problems are intended for use by either high school or university students. We use these problems in recitation sections in a beginning biology course at a university. Every week, each group of three to five students solves a problem related to that week's topic. Their solution is posted for other groups to compare.

In addition, the problems can be used for extra credit, club activities, filling unstructured time during a laboratory experiment, or for fostering competition or cooperation.

Many of the problems can be used for examples in mathematics classes. A great number of them (too many to list) involve the use of scientific notation and conversion of units. Here is a mathematical classification of many others:

- *Algebra:* Chapter 1: Problem 7; Chapter 5: Problem 8; Chapter 8: Problem 9; Chapter 12: Problem 3; Chapter 15: Problem 11; Chapter 16: Problems 5 and 6; Chapter 17: Problems 7 and 10; Chapter 18: Problem 3; Chapter 20: Problem 1; Chapter 21: Problems 22, 23, and 24; Chapter 22: Problem 6.

- *Geometry:* Chapter 4: Problems 4, 5, 6, 7, and 8; Chapter 5: Problems 4, 5, 6, 7, 8, 10, 11, and 12; Chapter 7: Problem 4; Chapter 8: Problems 6 and 9; Chapter 15: Problems 1, 6, 9, 10, and 11; Chapter 17: Problems 6, 7, 9, 10, and 12; Chapter 19: Problems 4, 11, 25, and 26: Chapter 20: Problems 1 and 5; Chapter 21: Problem 25; Chapter 22: Problem 2.

- *Graph interpretation:* Chapter 2: Problem 2; Chapter 3: Problems 7, 9, 10, and 11; Chapter 18: Problem 12.

- *Logarithms and exponentials:* Chapter 1: Problems 8 and 15; Chapter 2: Problems 3 and 5; Chapter 11: Problems 4, 5, and 6; Chapter 12: Problem 2; Chapter 14: Problem 9; Chapter 17: Problem 2; Chapter 18: Problem 18; Chapter 21: Problems 13 and 18; Chapter 22: Problems 3, 4, and 5.

- *Probability:* Chapter 11: Problems 4, 5, 6, and 7; Chapter 10: Problems 4, 5, and 6.

- *Statistics:* Chapter 10: Problems 1, 2, and 3; Chapter 14: Problem 5; Chapter 18: Problem 16; Chapter 22: Problem 9.

- *Trigonometry:* Chapter 21: Problems 1 and 3.

This book is a product of a grant from the National Science Foundation to the Indiana University System for the project Mathematics Throughout the Curriculum (grant number DUE-9555408). Additional funding came from the Indiana University Strategic Directions Initiative.
Comments and suggestions are welcome.

Robert Keck, rkeck@iupui.edu
Richard Patterson, rpatterson@math.iupui.edu

Contents

PART I CHEMISTRY

CHAPTER 1

Chemistry Review

Answers to Chapter 1 begin on page 179.

1-1 Below is a model of ATP. This molecule is the energy currency for most of life's processes.

Adenine

ATP Molecule

The P–O bonds marked with a wavy line are high-energy bonds that are more easily broken, allowing ATP to phosphorylate other molecules.

a. Compute the molecular weight of ATP:

Element	Number of Atoms		Atomic Weight		Molecular Weight
C	_____	×	_____	=	_____
H	_____	×	_____	=	_____
N	_____	×	_____	=	_____
P	_____	×	_____	=	_____
O	_____	×	_____	=	_____
Total molecular weight				=	_____

b. Below are listed all possible pairs of these five elements between which chemical bonds can exist. How many bonds are there of each type in one ATP molecule? (Count double bonds as two single bonds.)

c. Which possible bonds are not found in the ATP molecule?

d. What assumptions did you make?

Bond	Number	Bond	Number	Bond	Number
C–C	_____	H–H	_____	N–P	_____
C–H	_____	H–N	_____	N–O	_____
C–N	_____	H–P	_____	P–P	_____
C–P	_____	H–O	_____	P–O	_____
C–O	_____	N–N	_____	O–O	_____
P~O	_____				

1-2   Glucose (baker's sugar) can exist in two forms: a linear form and a ring form. The problem is to determine which of them has less energy, since that is the form that would normally be assumed.

Since the energy is accounted for by the chemical bonds, the total energy in a molecule can be computed by counting the chemical bonds of each type. Count the number of bonds of each type in both the linear and ring forms, and then use the table of bond energies to find the total energy in the linear form and the total energy in the ring form of glucose.

Linear

Ring

TABLE OF BOND ENERGIES

Bond	Energy (kcal/mol)
C–C	83
C–H	99
C–O	84
C=O	192
O–H	111
O=O	118

	LINEAR			RING		
Bond	Number of Bonds	Energy per Bond	Energy in Bonds	Number of Bonds	Energy per Bond	Energy in Bonds
C–C	_____	_____	_____	_____	_____	_____
C–H	_____	_____	_____	_____	_____	_____
C–O	_____	_____	_____	_____	_____	_____
C=O	_____	_____	_____	_____	_____	_____
O–H	_____	_____	_____	_____	_____	_____
Total energy:		Linear	_____		Ring	_____

a. Does the linear form or the ring form of glucose have less energy?

b. What is the percentage difference in energy between a double bond C=O and two single bonds C–O?

c. What assumptions did you make?

1-3 Suppose that the average weight of a biology student is 70 kg. Eighteen percent of this is carbon, 10% is hydrogen, 65% is oxygen, and 3% is nitrogen.

a. How much does the average biology student weigh in pounds? _____ lb

b. How much does the student's carbon weigh? _____ kg _____ lb

c. How much does the student's hydrogen weigh? _____ kg _____ lb

d. How much does the student's oxygen weigh? _____ kg _____ lb

e. How much does the student's nitrogen weigh? _____ kg _____ lb

f. What assumptions did you make?

1-4 Table salt (NaCl) makes up 0.2% of the weight of a 70 kg biology student.

a. What is the weight, in kilograms and in pounds, of table salt in the biology student?

b. How many students are needed to have at least as much as a one-pound box of salt?

c. What assumptions did you make?

1-5 The diameter of an atom is about 4 orders of magnitude (1×10^4) larger than the diameter of the nucleus itself. A table tennis ball is about $1\frac{7}{16}$ inches in diameter.

a. Using the table tennis ball as a representation of the nucleus, how far away from the table tennis ball would electrons be found?

b. Which of the following distances best approximates the answer?

　i. The length of your hand

　ii. The length of a football field

　iii. The width of Indiana

　iv. The distance from New York to Los Angeles

c. What assumptions did you make?

1-6 The glucose content in blood is in the range of 50 milligrams per deciliter to 300 mg/dl. Using a value of 3 liters of blood for the average adult human, what is the range of total available glucose moving in the bloodstream?

	At 50 mg/dl	At 300 mg/dl
a. In grams	_____	_____
b. In moles	_____	_____
c. In number of molecules	_____	_____

d. What assumptions did you make?

1-7 Some of the mass of your body is composed of protons, some of neutrons, and some of electrons. Recall that an electron weighs 1/2000 as much as a nuclear particle.

a. What is a reasonable estimate of the weight of the positively charged particles in a 70 kg biology student?

b. What is a reasonable estimate of the weight of the electrically neutral particles?

c. What is a reasonable estimate of the weight of the negatively charged particles?

d. What assumptions did you make?

1-8 Some isotopes of atoms spontaneously undergo radioactive decay. In radioactive decay, the number of neutrons and/or protons in the nucleus decreases and the molecule emits energy. The energy can be measured by a counter, such as a Geiger counter, that registers in counts per minute (cpm) the decay events.

^{32}P is a radioactive isotope that has a half-life of 14.3 days, which means that only 50% of the original amount will remain at that time.

An experiment is conducted in which ^{32}P is incorporated into developing DNA strands. A counter registers initially 10^8 cpm per μg of DNA.

The sample is saved in order to compare it to another one that is generated the following week. Then all of the samples must be stored for at least 6 half-lives before they can be disposed of safely.

a. What is the decay rate of ^{32}P?

b. What will be the cpm exactly 7 days later?

c. If the samples are stored in a freezer for one year prior to disposal, is this a safe enough length of time?

Chapter 1

d. What will be the cpm in the stored sample after one year?

e. What assumptions did you make?

1-9   Water molecules are *dipolar* because of partially charged positive and negative ends. Consequently, they are attracted to one another and other ions in solution. Ions, because they have a stronger charge, tend to cause water molecules to adhere more tightly to them than to other water molecules. On average, four water molecules typically bind to each ion and are removed from the more free water molecules. This lowers water's ability to freeze or boil or dissolve other materials.

a. What percentage of the water is immobilized in a plant cell with an internal KCl solution equal to 150 mM?

b. What assumptions did you make?

1-10   Glycogen is a polymer of glucose that is branched approximately every 10 glucose molecules. The oval encircles one repeating subunit.

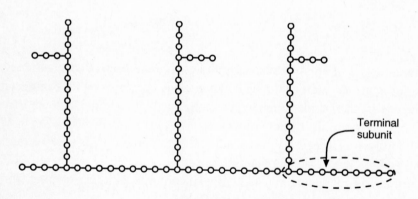

Glycogen

Up to 10% of the weight of the liver is composed of glycogen, which is mobilized into the bloodstream to maintain glucose concentrations within a physiological range. Liver mass averages 1500 g in human adults. Assume that all of the stored glucose in an individual is mobilized into the bloodstream over a period of several days.

a. Consider a terminal subunit of glycogen, consisting of 10 glucose units. How many water molecules are utilized to break the bonds to release these 10 glucose units?

b. How many grams of glucose were released by the liver?

c. The brain of an adult uses about 140 g of glucose each day. How many hours of "brain food" are stored in the liver?

d. What assumptions did you make?

1-11   Assume that the average male biology student weighs 165 pounds. Mature males average 60% water, females about 50% water. Cell water (water inside the cell membrane) makes up about 2/3 of the total volume. The other 1/3 is about 3/4 interstitial and 1/4 blood water (plasma).

a. How many liters of water are present in an average male biology student?

b. How many molecules of water are present in the same student?

c. What is the volume of plasma in the male student?

d. What assumptions did you make?

1-12   A farmer cuts a field of alfalfa to make hay. When growing, the alfalfa is 80% water by weight. After it is cut, it must be allowed to dry out before it is baled because baled hay will mold if it contains more than 15% water. A good hay crop will yield 100 bales per acre, weighing an average of 85 pounds each.

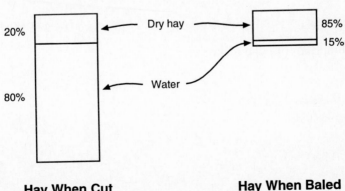

Hay When Cut   **Hay When Baled**

a. What is the weight (in pounds) of water lost per bale from the time of cutting the hay to baling it?

b. What is the weight (in pounds) of water lost per acre?

c. What is the weight (in kilograms) of water lost per bale?

d. What is the volume (in liters) of water lost per bale?

e. What is the volume (in quarts) of water lost per bale?

f. What is the volume (in quarts) of water baled per acre?

g. What assumptions did you make?

1-13 The isotope ^{12}C is a stable isotope of carbon. ^{14}C is an unstable isotope of carbon.

a. How much heavier is ^{14}C compared to ^{12}C?

b. If the C in glucose ($C_6H_{12}O_6$) were all ^{14}C, how much heavier would this molecule be when compared to ^{12}C glucose?

c. What assumptions did you make?

1-14 The isotope ^{32}P is radioactive but ^{31}P is not. ^{32}P in phosphate is used as a "marker" in intact plants and animals as well as a marker in laboratory experiments when searching for genes. In particular, ATP can be made in the laboratory using ^{32}P.

a. How much heavier is the phosphate ion containing ^{32}P compared to the phosphate ion with ^{31}P?

b. How much heavier is ATP that contains ^{32}P as compared to ATP with ^{31}P?

c. What assumptions did you make?

Adenine

ATP Molecule

1-15 The smallest volume the stomach can have is about 25 milliliters, with the pH of the contents at 0.84. The largest it can have is about 2 liters. Assume you exercise vigorously all day and then drink 2 liters of water. What is the pH of the stomach contents after you drink the water?

a. What is the hydrogen ion concentration of the stomach at its smallest?

b. How many hydrogen ions are there in an empty stomach?

c. How many hydrogen ions are there in 2 liters of water?

d. How many in total in the full stomach?

e. What is the new concentration?

f. What is the pH of the stomach contents after you drink the water?

g. List any assumptions you made.

Biomolecules

Answers to Chapter 2 begin on page 186.

2-1 Many globular proteins contain a segment of amino acids that is extended (stretched out) and after making a U-turn binds with itself. This is one kind of β pleated sheet known as *antiparallel*. Instead of H bonds assisting in forming a spiral (as in an α helix), H bonds exist between adjacent amino acids so as to cause a ladder-like arrangement stabilizing that part of the protein.

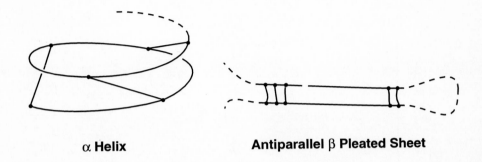

α Helix   **Antiparallel β Pleated Sheet**

a. In an antiparallel β pleated sheet containing 100 amino acids, 50 in each strand, how many H bonds would form?

b. In an α helix of 50 amino acids, how many H bonds would form?

c. What assumptions did you make?

2-2 Amino acids play a role in buffering pH changes in the cytoplasm of a cell. They do this by changing from an un-ionized form into a dipolar ion with two opposing ionic regions called a *zwitterion*. Here is the zwitterion form of alanine:

$$H_3N^+ - \overset{\displaystyle H}{\underset{\displaystyle CH_3}{C}} - C \overset{\displaystyle O}{\underset{\displaystyle O^-}{\diagup}}$$

The Zwitterion Form of Alanine

In very acidic conditions, an extra proton attaches at the O^-. In very basic conditions, both this proton and one from H_3N^+ are lost.

Chapter 2

Alanine has the following *titration curve*, relating the pH of the cell to the concentration of OH^-. The pH is labeled at three points on the graph. Notice that near several of these points the cell has a *bufferering capacity*; that is, it can absorb OH^- without causing much change in pH. At pH = 2.34, half the protons are lost from O^- and water is formed. At pH = 9.69, half the protons are lost from H_3N^+ and again water is formed.

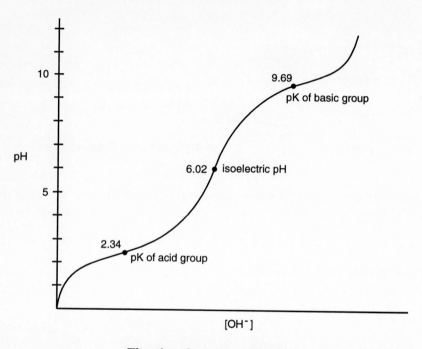

Titration Curve for Alanine

Use the information in the graph to compare the sizes of each of the following pairs. In each blank write <, >, or = according to whether the first quantity is less than, greater than, or the same as the second quantity, and explain.

a. The number of negatively charged molecules: at pH = 1 _____ at pH = 12.

b. The rate of change in pH: at pH = 2.34 _____ at pH = 9.69.

c. The rate of change in pH: at pH = 6.02 _____ at pH = 9.69.

d. The percentage of zwitterion: at pH = 6.02 _____ at pH = 9.69.

e. The percentage of zwitterion: at pH = 2.34 _____ at pH = 9.69.

f. The buffering capacity: at pH = 6.02 _____ at pH = 9.69.

g. The ratio of positively charged molecules to neutral molecules: at pH $= 2.34$ _____ at pH $= 9.69$.

h. The ratio of positively charged molecules to neutral molecules: at pH $= 6.02$ _____ at pH $= 9.69$.

i. What assumptions did you make?

2-3 Proteins (both enzymatic and structural) play a significant role in maintaining the pH in a cell. Seven of the 20 amino acids commonly found in proteins have an ionizable group. These groups (side chains) determine the charge on the protein and buffer the cytoplasm of the cell that contains it.

Populations of proteins that contain ionizable side chains have a probability of being "charged" or not. The probability of a side chain being charged is related primarily to the side chain chemistry and the pH of the cellular compartment; other usually less important factors are temperature, ionic strength, and bulk microenvironments.

The ratio of the charged to uncharged side chains is usually symbolized by K:

$$K = \frac{c}{u}$$

where c is the proportion of charged side chains and u is the proportion of uncharged side chains (so that $c + u = 1$).

In the laboratory, the pH of a protein's environment can be altered so that we have equal numbers of charged and uncharged molecules. We call the pH at which K is 1 (equal number of charged and uncharged side chains) the pK of the side chain.

Below is a list of five amino acids whose charge is altered by the cellular pH:

Acidic	pK	Basic	pK
Aspartic acid	3.65	Histidine	6.00
Glutamic acid	4.25	Lysine	10.53
		Arginine	12.48

Cell compartments can vary in pH from about 4.0 to 8.5, but the pK values of the various amino acids remain fixed. When the pH of the environment is not equal to the pK of the side chain, the proportion of charged and uncharged side chains changes so as to satisfy the equation

$$pH = pK + \log \frac{c}{u}$$

As an example, for aspartic acid in a protein with a pH environment of 3.87, $pH = pK$ so $\log \frac{c}{u} = 0$. Consequently $\frac{c}{u} = 10^0 = 1$, and $c = u$. However, when the pH and the pK are different, the term $\log \frac{c}{u}$ will not be zero and the ratio $\frac{c}{u}$ will be different from 1.

a. What percentage of the aspartic acid molecules are charged at $pH = 4.0$?

b. What percentage of the aspartic acid molecules are uncharged at $pH = 4.0$?

c. What percentage of the aspartic acid molecules are charged at $pH = 8.5$?

d. What percentage of the aspartic acid molecules are uncharged at $pH = 8.5$?

e. What assumptions did you make?

2-4 The sequence of amino acids of the enzyme lysozyme is known. Below is a list of amino acids and the number of each in the lysozyme molecule.

Type	Molecular Weight	Number in Lysozyme
Alanine	89	12
Arginine	174	11
Asparagine	132	13
Aspartic acid	133	8
Cysteine	121	8
Glutamic acid	147	2
Glutamine	146	3
Glycine	75	12
Histidine	155	1
Isoleucine	131	6
Leucine	131	8
Lysine	146	6
Methionine	149	2
Phenylalanine	165	3
Proline	115	2
Serine	105	10
Threonine	119	7
Tryptophan	204	6
Tyrosine	181	3
Valine	117	6

a. What is the average molecular weight of the 20 amino acids?

b. What is the molecular weight of lysozyme?

c. What is the average molecular weight of the amino acids in lysozyme?

d. How many S-S bonds are possible in lysozyme?

e. Is the net charge on lysozyme positive or negative?

f. What assumptions did you make?

2-5 Proteins (both enzymatic and structural) play a significant role in maintaining the pH in a cell. Seven of the 20 amino acids commonly found in proteins have an ionizable group. These groups (side chains) determine the charge on the protein and buffer the cytoplasm of the cell that contains it.

Populations of proteins that contain ionizable side chains have a probability of being "charged" or not. The probability of a side chain being charged is related primarily to the side chain chemistry and the pH of the cellular compartment; other usually less important factors are temperature, ionic strength, and bulk microenvironments.

The ratio of the charged to uncharged side chains is usually symbolized by K:

$$K = \frac{c}{u}$$

where c is the proportion of charged side chains and u is the proportion of uncharged side chains (so that $c + u = 1$).

In the laboratory, the pH of a protein's environment can be altered so that we have equal numbers of charged and uncharged molecules. We call the pH at which K is 1 (equal number of charged and uncharged side chains) the pK of the side chain.

Below is a list of the seven amino acids whose charge is altered by the cellular pH:

Acidic	pK	Basic	pK	Polar	pK
Aspartic acid	3.87	Lysine	10.53	Cysteine	8.33
Glutamic acid	4.25	Arginine	12.48	Tyrosine	10.07
		Histidine	6.00		

Cell compartments can vary in pH from about 4.0 to 8.5, but the pK values of the various amino acids remain fixed. When the pH of the environment is not equal to the pK of the side chain, the proportion of charged and uncharged side chains changes so as to satisfy the equation

$$pH = pK + \log \frac{c}{u}$$

As an example, for aspartic acid in a protein with a pH environment of 3.87, pH = pK so $\log \frac{c}{u} = 0$. Consequently $\frac{c}{u} = 10^0 = 1$, and $c = u$. However, when the pH and the pK are different, the term $\log \frac{c}{u}$ will not be zero and the ratio $\frac{c}{u}$ will be different from 1.

The pH of healthy cell cytoplasm varies from 7.2 to 7.4. Using this information and the equation above, complete the following table:

Amino Acid in Protein	pK of the Side Chain	% Charged at pH = 7.2	% Charged at pH = 7.4
Aspartic acid	3.87	_____	_____
Histidine	6.0	_____	_____
Cysteine	8.33	_____	_____
Arginine	12.48	_____	_____

What assumptions did you make?

2-6 Polymers of glucose units are used as temporary food storage in both plant and animal cells. Glucose units are connected to one another by 1,4-linkages to make a linear polymer and by 1,6-linkages to make branch points.

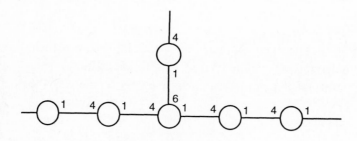

Polymer Linkages

Polysaccharides of glucose units vary in size. The three most commonly encountered are

Type of Starch	Cell Type	Polymer Size	Averge Number of 1,4-Bonds Between Branches
Amylopectin	plant	100,000,000	24 to 30
Amylose	plant	500,000	linear
Glycogen	animal	3,000,000	8 to 12

When each polymer bond is made, a water molecule is released and becomes part of the cell water.

a. How many water molecules were released during formation of each of the three polysaccharides?

    If plant cells are about 90% water and the dimensions of a plant cell are 18μm by 18μm by 60μm, and an animal cell is also 90% water and is a sphere with a diameter of $25\ \mu$m:

b. What is the percentage increase in cell water when one starch molecule of each type is made?

c. What assumptions did you make?

CHAPTER 3

Enzymatics
and Energetics

Answers to Chapter 3 begin on page 191.

3-1 Respiration requires many enzymes; two respiratory enzymes and their maximum rates of reaction are given below:

Enzyme	V_{max}
Catalase	5,600,000
Succinate dehydrogenase	1,150

where V_{max} is measured in molecules of product formed (or substrate utilized) per molecule of enzyme per minute.

a. How many molecules of substrate can be utilized in one second by catalase?

b. How many molecules of substrate can be utilized in one second by dehydrogenase?

c. How many times faster is catalase at producing product compared to dehydrogenase?

d. What assumptions did you make?

3-2 One carbonic anhydrase molecule can catalyze 36,000,000 substrates to products in one minute.

a. How much time does it take for carbonic anhydrase to attach to one substrate molecule, convert substrate to products, and release the products?

b. What assumptions did you make?

3-3 The symbol ΔG stands for the change in free energy during a chemical reaction. If the concentration of both products and reactants is 1 mole per liter, then the symbol ΔG^o is used, and if in addition the pH is 7, then $\Delta G^{o'}$ is used.
ΔG can be calculated from $\Delta G^{o'}$ by means of the formula

$$\Delta G = \Delta G^{o'} + RT \ln \frac{[\text{Product}_1][\text{Product}_2] \cdots [\text{Product}_m]}{[\text{Reactant}_1][\text{Reactant}_2] \cdots [\text{Reactant}_n]}$$

if logarithms base e are used, or by the formula

$$\Delta G = \Delta G^{o'} + 2.303 RT \log \frac{[\text{Product}_1][\text{Product}_2] \cdots [\text{Product}_m]}{[\text{Reactant}_1][\text{Reactant}_2] \cdots [\text{Reactant}_n]}$$

Chapter 3

if logarithms base 10 are used. Here m is the number of products, n is the number of reactants, R is the universal gas constant, $R = 1.98 \times 10^{-3}$ kilocalories per mole per degree Kelvin, and T is the temperature in degrees Kelvin. The concentrations are measured in moles per liter, but for technical reasons, the units are not used with them.

For the particular reaction

$$\text{ATP} \longleftrightarrow \text{ADP} + \text{P}_i,$$

$$\Delta G = \Delta G^{o'} + RT \ln \frac{[\text{ADP}][\text{P}_i]}{[\text{ATP}]}$$

Here $\Delta G^{o'} = -7.3$ kcal/mol at $T = 298 (= 273 + 25°C)$.

Under physiological concentrations, the energy yield from the hydrolysis of ATP can be quite different.

The concentrations in the cytoplasm of a plant cell are

Chemical	Concentration
ATP	3 mM
ADP	1 mM
P_i	10 mM

a. What would be the value of ΔG of ATP hydrolysis in this plant cell?

The concentrations in the cytoplasm of a muscle cell after exercise are

Chemical	Concentration
ATP	4 mM
ADP	0.013 mM
P_i	140 mM

b. What would be the value of ΔG of ATP hydrolysis in this animal cell?

c. What assumptions did you make?

3-4   The symbol ΔG stands for the change in free energy during a chemical reaction. If the concentration of both products and reactants is 1 mole per liter, then the symbol ΔG^{o} is used, and if in addition the pH is 7, then $\Delta G^{o'}$ is used.

ΔG can be calculated from $\Delta G^{o'}$ by means of the formula

$$\Delta G = \Delta G^{o'} + RT \ln \frac{[\text{Product}_1][\text{Product}_2] \cdots [\text{Product}_m]}{[\text{Reactant}_1][\text{Reactant}_2] \cdots [\text{Reactant}_n]}$$

if logarithms base e are used, or by the formula

$$\Delta G = \Delta G^{o'} + 2.303\,RT \log \frac{[\text{Product}_1][\text{Product}_2] \cdots [\text{Product}_m]}{[\text{Reactant}_1][\text{Reactant}_2] \cdots [\text{Reactant}_n]}$$

if logarithms base 10 are used. Here m is the number of products, n is the number of reactants, R is the universal gas constant, $R = 1.98 \times 10^{-3}$ kilocalories per mole per degree Kelvin, and T is the temperature in degrees Kelvin. The concentrations are measured in moles per liter, but for technical reasons, the units are not used with them.

For the particular reaction

$$\text{ATP} \longleftrightarrow \text{ADP} + \text{P}_i,$$

$$\Delta G = \Delta G^{o'} + RT \ln \frac{[\text{ADP}][\text{P}_i]}{[\text{ATP}]}$$

Here $\Delta G^{o'} = -7.3$ kcal/mol at $T = 298 (= 273 + 25°\text{C})$.

The temperature of a leaf varies due to the environmental temperature. A range of temperature during the summer in the midwestern United States is from $20°$ to $35°\text{C}$.

The concentrations in the cytoplasm of the leaf cell are

Chemical	Concentration
ATP	3 mM
ADP	1 mM
P_i	10 mM

a. Which temperature causes the larger ΔG from ATP hydrolysis?

b. What percentage increase is realized due to the temperature range from minimum to maximum?

c. What assumptions did you make?

3-5   The symbol ΔG stands for the change in free energy during a chemical reaction. If the concentration of both products and reactants is 1 mole per liter, then the symbol ΔG^{o} is used, and if in addition the pH is 7, then $\Delta G^{o'}$ is used.

ΔG can be calculated from $\Delta G^{o'}$ by means of the formula

$$\Delta G = \Delta G^{o'} + RT \ln \frac{[\text{Product}_1][\text{Product}_2] \cdots [\text{Product}_m]}{[\text{Reactant}_1][\text{Reactant}_2] \cdots [\text{Reactant}_n]}$$

Chapter 3

if logarithms base e are used, or by the formula

$$\Delta G = \Delta G^{o'} + 2.303RT \log \frac{[\text{Product}_1][\text{Product}_2]\cdots[\text{Product}_m]}{[\text{Reactant}_1][\text{Reactant}_2]\cdots[\text{Reactant}_n]}$$

if logarithms base 10 are used. Here m is the number of products, n is the number of reactants, R is the universal gas constant, $R = 1.98 \times 10^{-3}$ kilocalories per mole per degree Kelvin, and T is the temperature in degrees Kelvin. The concentrations are measured in moles per liter, but for technical reasons, the units are not used with them.

For the particular reaction

$$\text{ATP} \longleftrightarrow \text{ADP} + \text{P}_i,$$

$$\Delta G = \Delta G^{o'} + RT \ln \frac{[\text{ADP}][\text{P}_i]}{[\text{ATP}]}$$

Here $\Delta G^{o'} = -7.3$ kcal/mol at $T = 298 (= 273 + 25°C)$.

In different types of animal cells, the concentrations of products and reactants vary. This alters the energy realized by ATP hydrolysis. In a rat, for example, with body temperature averaging 101°F, three different types of cells have the following concentrations:

Cell Type	ATP	ADP	P_i
Hepatocyte	3.38 mM	1.32 mM	4.8 mM
Myocyte	8.05 mM	0.93 mM	8.05 mM
Neuron	2.59 mM	0.73 mM	2.72 mM

a. What is the average body temperature of a rat in degrees Kelvin?

b. Which type of rat cell has the most negative ΔG when ATP hydrolyzes?

c. What is the energy difference in the rat cells with the highest and lowest ΔG?

d. What assumptions did you make?

3-6 Hexokinase is an enzyme that adds phosphate to a glucose molecule, one of its possible substrates. This makes the glucose molecule more reactive. During the reaction, the 3-D structure of hexokinase is altered. This change in structure causes P_i to be added to the glucose and causes 100 molecules of water to be lost from the protein part of the enzyme.

a. What is the ratio of the mass of water "lost" by the enzyme to the mass of phosphate "gained" by the glucose when the substrate is added?

b. Does the behavior of hexokinase give evidence for the lock-and-key theory or the induced-fit theory of enzymes?

c. What assumptions did you make?

3-7 Hydrogen peroxide is usually stored in a brown bottle away from sunlight because it spontaneously (but slowly) decomposes into O_2 and water. In the brown bottle, the free energy of activation $\Delta G^+ = 18$ kcal/mol. In the presence of a catalyst, the decomposition is much faster. For each decrease of 1.36 kcal/mol in ΔG^+, the rate of reaction is ten times faster.

In the presence of catalase, an enzyme found in the blood, $\Delta G^+ = 7$ kcal/mol. In the presence of the inorganic catalyst platinum, $\Delta G^+ = 13$ kcal/mol.

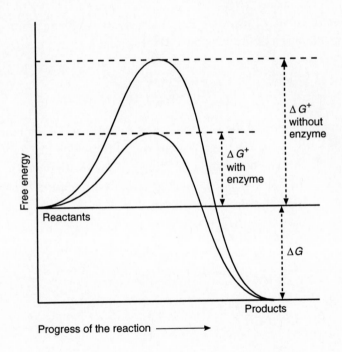

a. How much faster can catalase cause peroxide decomposition?

b. How much faster can platinum cause peroxide decomposition?

c. How many times faster is peroxide degradation under the influence of catalase than platinum degradation?

d. What assumptions did you make?

3-8 Lactate dehydrogenase in rat muscle is an enzyme with a molecular weight of 134 kilodaltons. An assay is made of the enzyme in order to compare its activity before and after it is inhibited.

A molecule of the enzyme is composed of four identical subunits, each of which is believed to have one active site. A substance is added that is known to be an irreversible inhibitor acting at the active site.

Suppose 1 mg of lactate dehydrogenase is used for each enzyme assay.

a. How many inhibitor molecules would be needed to tie up half of the active sites in the assay?

b. What assumptions did you make?

3-9 The *Michaelis-Menton* function relates the velocity of the reaction caused by an enzyme to the concentration of the substrate. The formula is

$$V = \frac{V_{max} \cdot [S]}{K_m + [S]}$$

where $[S]$ is the concentration of the substrate, V is the velocity of the reaction, V_{max} is the maximum possible velocity, and K_m is the value of $[S]$ for which the velocity is one-half V_{max}. The graph of the Michaelis-Menton function is drawn with the concentration $[S]$ of substrate measured on the horizontal axis and the velocity V measured on the vertical axis; see Figure (a). The curve is a segment of a hyperbola, with $V = V_{max}$ as one of its asymptotes.

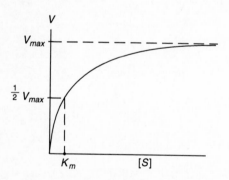

(a) The Michaelis-Menton Graph

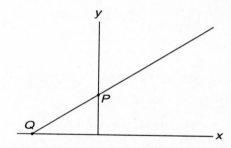

(b) The Lineweaver-Burk Graph

By algebraic manipulation, this equation is equivalent to

$$\frac{1}{V} = \frac{K_m}{V_{max}} \frac{1}{[S]} + \frac{1}{V_{max}}$$

Its graph is the straight line

$$y = mx + b$$

if the variables are taken to be $x = 1/[S]$ on the horizontal axis and $y = 1/V$ on the vertical axis, the slope $m = K_m/V_{max}$, and the y-intercept $b = 1/V_{max}$; see Figure (b). It is called the *Lineweaver-Burk* graph of the data.

a. Find the coordinates of the points P and Q where the Lineweaver-Burk graph meets the axes and the slope of the line in terms of K_m and V_{max}.

In a scientific paper you read that an enzyme you wish to study has $K_m = 0.1$ mM and $V_{max} = 2500$ mg product made per mg protein per minute. You repeat the published experiments in your laboratory and confirm their results.

You have a mutant cell line that you believe is "better" at producing product. Because the product is economically important, it is imperative that you compare your rate of product formation against the published rate. Both enzymes exhibit Michaelis-Menton kinetics.

b. Construct the Lineweaver-Burk graph of the enzyme. (Graph paper is provided on the next page.)

c. Similarly, graph the Michaelis-Menton rectangular hyperbola for this data.

Now you measure your "mutant" enzyme and find that $V_{max} = 2400$ and K_m is 25% smaller.

d. Construct the Lineweaver-Burk graph for the mutant enzyme.

e. Construct the Michaelis-Menton graph for the mutant enzyme.

f. Which enzyme is better and why?

g. What assumptions did you make?

Chapter 3

3-10 The *Michaelis-Menton* function relates the velocity of the reaction caused by an enzyme to the concentration of the substrate. The formula is

$$V = \frac{V_{max} \cdot [S]}{K_m + [S]}$$

where $[S]$ is the concentration of the substrate, V is the velocity of the reaction, V_{max} is the maximum possible velocity, and K_m is the value of $[S]$ for which the velocity is one-half V_{max}. The graph of the Michaelis-Menton function is drawn with the concentration $[S]$ of substrate measured on the horizontal axis and the velocity V measured on the vertical axis; see Figure (a). The curve is a segment of a hyperbola, with $V = V_{max}$ as one of its asymptotes.

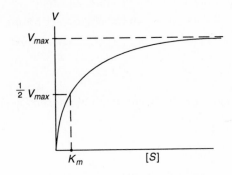

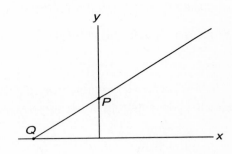

(a) The Michaelis-Menton Graph   **(b) The Lineweaver-Burk Graph**

By algebraic manipulation, this equation is equivalent to

$$\frac{1}{V} = \frac{K_m}{V_{max}} \frac{1}{[S]} + \frac{1}{V_{max}}$$

Its graph is the straight line

$$y = mx + b$$

if the variables are taken to be $x = 1/[S]$ on the horizontal axis and $y = 1/V$ on the vertical axis, the slope $m = K_m/V_{max}$, and the y-intercept $b = 1/V_{max}$; see Figure (b). It is called the *Lineweaver-Burk* graph of the data.

  a. Find the coordinates of the points P and Q where the Lineweaver-Burk graph meets the axes and the slope of the line in terms of K_m and V_{max}.

The table below gives the values of K_m and V_{max} for five enzymes. K_m is measured in micromolars and V_{max} in milligrams of product made per milligrams of enzyme per minute.

Enzyme	K_m	V_{max}	Slope
Carbonic anhydrase	9,000	36,000,000	_____
Catalase	25,000	5,600,000	_____
Penicillinase	50	120,000	_____
Chymotrypsin	5,000	6,000	_____
Lysozyme	6	30	_____

b. Fill in the the slope of this line for each of the enzymes in the table.

c. Choose one of the enzymes and draw both the Lineweaver-Burk and the Michaelis-Menton graph for it. Mark the location of K_m in the Michaelis-Menton graph.

d. What assumptions did you make?

3-11 The *Michaelis-Menton* function relates the velocity of the reaction caused by an enzyme to the concentration of the substrate. The formula is

$$V = \frac{V_{max} \cdot [S]}{K_m + [S]}$$

where $[S]$ is the concentration of the substrate, V is the velocity of the reaction, V_{max} is the maximum possible velocity, and K_m is the value of $[S]$ for which the velocity is one-half V_{max}. The graph of the Michaelis-Menton function is drawn with the concentration $[S]$ of substrate measured on the horizontal axis and the velocity V measured on the vertical axis; see Figure (a). The curve is a segment of a hyperbola, with $V = V_{max}$ as one of its asymptotes.

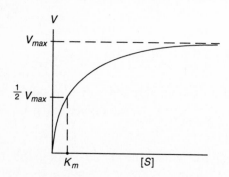

(a) The Michaelis-Menton Graph

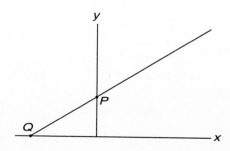

(b) The Lineweaver-Burk Graph

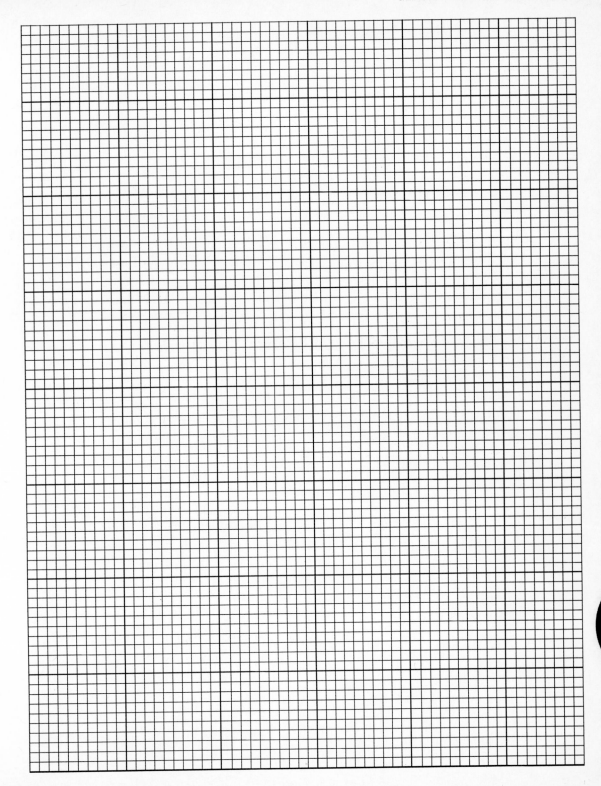

By algebraic manipulation, this equation is equivalent to

$$\frac{1}{V} = \frac{K_m}{V_{max}} \frac{1}{[S]} + \frac{1}{V_{max}}$$

Its graph is the straight line

$$y = mx + b$$

if the variables are taken to be $x = 1/[S]$ on the horizontal axis and $y = 1/V$ on the vertical axis, the slope $m = K_m/V_{max}$, and the y-intercept $b = 1/V_{max}$; see Figure (b). It is called the *Lineweaver-Burk* graph of the data.

a. Find the coordinates of the points P and Q where the Lineweaver-Burk graph meets the axes and the slope of the line in terms of K_m and V_{max}.

   For the enzyme catalase, $K_m = 25$ mM and $V_{max} = 5,600,000$ molecules of product per molecule of enzyme per minute.

b. Draw the Lineweaver-Burk graph for catalase. (The units on the two axes may need to be different.)

c. Draw the Michaelis-Menten graph for catalase.

d. Substitute $[S] = K_m$ into the equation for V and simplify it, to confirm that the enzymatic rate V at a substrate concentration equal to K_m is indeed $\frac{1}{2}V_{max}$.

e. What is the enzymatic rate at a substrate concentration equal to $\frac{1}{2}K_m$?

f. What is the enzymatic rate at a substrate concentration equal to $4K_m$?

g. Plot the points obtained from (d), (e), and (f) on your graph (c).

h. What assumptions did you make?

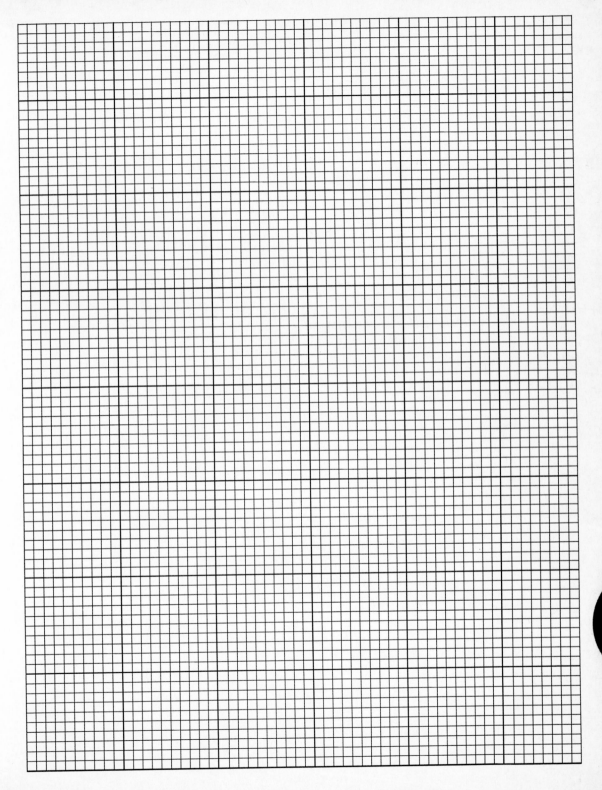

PART II CELLS

Answers to Chapter 4 begin on page 197.

CHAPTER 4

Cell Structure

4-1 Human nerve cells can transport mitochondria from the bulk of the cellular contents down a long narrow tube of cytoplasm called the *axon*. This transport uses ATP and is guided along microtubules. The axon may be as long as 1 meter and the rate of mitochondrial movement is 0.6 micrometer per second.

a. How long will it take to move a mitochondrion 1 meter?

b. What assumptions did you make?

4-2 Some cells in humans crawl all the time; these crawling cells assist in wound healing, blood clot formation, infection fighting, and also in cancer spreading. Some cells crawl more slowly than others. Wound-healing cells and cancerous cells crawl at a rate from 0.1 to 1 micrometer per hour. Infection fighters in the lungs and gastrointestinal tract can crawl at a maximal rate of 30 micrometers per minute.

a. How much faster can infection fighters crawl when compared to wound-healing and cancerous cells?

b. What assumptions did you make?

4-3 Some of the longest microtubules in nonspecialized cells are 10 to 25 μm long. Microtubules are often composed of 13 *protofilaments* adjacent to one another along their long axes. The filaments are composed of heterodimers called *tubulin subunits*. Each heterodimer is about 8 nm long.

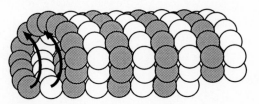

Microtubule

Tubulin protofilament

Tubulin heterodimer subunit

Chapter 4

a. How many tubulin subunits are required to make a microtubule 20 μm long?

b. What assumptions did you make?

4-4 *Clathrin* is a fibrous protein that exists in the shape of a triskelion; see Figure (a). The triskelions polymerize by joining together to form a curved polyhedral structure made up of 8 hexagons and 12 pentagons; see Figure (b). In all, 36 triskelions make up the structure, with one of them centered at each of the 36 vertices. These structures cause the invagination ("pinching in") of the cell membrane to make coated vesicles. Thus, they function in the process of endocytosis.

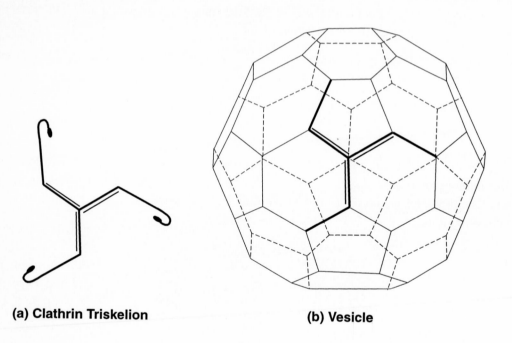

(a) Clathrin Triskelion   **(b) Vesicle**

The vesicles are variable in size but average about 40 nm in diameter. After vesiculation the triskelions are recycled to generate additional vesicles. Triskelion recycling requires ATP. One cultured animal cell (fibroblast) can generate up to 2500 vesicles each minute.

a. How many clathrin molecules are required to form these 2500 vesicles? Assume one ATP is needed for each triskelion recycling.

b. How many ATP molecules are needed for recycling these 2500 vesicles?

c. How much cell surface is made into vesicles each minute?

d. What assumptions did you make?

4-5 The cytoskeleton of a red blood cell (RBC) is partially composed of protein molecules called *spectrin*. Each unit of spectrin is composed of two proteins (alpha, with molecular weight 260 kilodaltons, and beta, with molecular weight 225 kd) running in a parallel twisted cable. Two such units combine head to tail to yield a tetramere 200 nanometers long. The 200 nm protein complex attaches to proteins embedded in the cell membrane. A red blood cell during development can be modeled as a sphere about 6 μm in diameter.

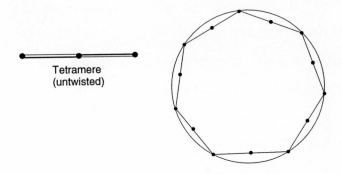

Tetramere
(untwisted)

Red Blood Cell

a. How many 200 nm complexes would be needed to go once around the inner surface of the developing RBC?

b. What is the minimal number of spectrin attachment sites for once around the cell?

c. What is the total molecular weight of this protein cable that goes around the inner (cytoplasmic) surface once?

d. Assume the average amino acid to have molecular weight 110 daltons. How many linked amino acids are needed?

e. What assumptions did you make?

Chapter 4

4-6 *Lysosomes* are little sacs (spheres) of acid in a cell. Their pH is about 5 and an electron micrograph suggests they have a diameter of 0.5 μm. The increased hydrogen ion concentration is due to the pumping of hydrogen ions across the membrane from a region with a pH of 7.2.

    a. Assuming no buffering capacity inside the lysosome, how many hydrogen ions were moved to the inside?

    b. What assumptions did you make?

4-7 Lipid droplets can be seen in electron micrographs of most cells; some cells known as *adipocytes* are specially adapted to store fats. Spherical lipid droplets vary in size but can be large (1.2 μm in diameter). If the stored fat is a triglyceride (tripalmitate), which is 9/10 as dense as water, what is the number of triglyceride molecules in this fat body?

    a. What is the volume of the fat body?

    b. What would be the weight of water with exactly the same volume as the fat body?

    c. What is the weight of the lipid in this fat body?

    d. What is the molecular weight of tripalmitate ($C_{51}H_{98}O_6$)?

    e. What is the number of triglyceride molecules in this fat body?

    f. What assumptions did you make?

4-8 *Microvilli* are extensions of cells of the small intestine epithelium. Microvilli are rodlike (1 μm long and 0.1 μm diameter) and are numerous (2 billion per square centimeter of intestinal epithelium).

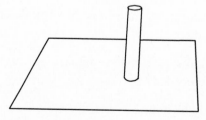

    a. What is the surface area of one microvillus?

    b. What is the surface area of the microvilli per square centimeter?

    c. By what percentage is the absorptive intestinal surface increased by these microvilli?

    d. What assumptions did you make?

4-9   By weight, primary plant cell walls are about 90% polysaccharide and 10% glycoprotein. In dicot cells, the polysaccharide component is one-third cellulose (repeating glucose units), one-third hemicellulose (repeating units with branches of five-carbon sugars), and one-third pectins (repeating six-carbon sugar acids). The glycoprotein component is about 35% protein and 65% carbohydrate. The amino acid composition of the protein has repeating units of lysine, glutamic acid, and tyrosine.

    a. What percentage of the cell wall is protein?

    b. Which type polymer determines the pH of the cell wall?

    c. Would you expect the cell wall to be more acidic or more basic?

    d. What assumptions did you make?

CHAPTER 5

Membranes

Answers to Chapter 5 begin on page 201.

5-1 The protein that facilitates glucose entry into human cells is composed of 492 amino acids. The protein traverses the lipid bilayer 12 times and each span requires 21 amino acids. Both the C end of this protein and the N end of the protein are in the cytoplasm.

a. What percentage of the amino acids are involved in spanning the membrane?

b. How many non-membrane-spanning segments of the protein are there?

c. What is the average number of amino acids in each non-membrane-spanning part of this protein?

d. What assumptions did you make?

5-2 Cholesterol is removed from the bloodstream by cells that make the protein LDL (low-density lipoprotein), each molecule of which can bind about 2000 cholesterol molecules. The LDL is brought into a cell in a vesicle generated by endocytosis. The LDL is recycled to the membrane surface by transport vesicles so that one round trip of an LDL (into and out of the cell) takes 10 minutes. LDLs usually have a life span of 20 hours.

a. How many cholesterol molecules can be brought into the cell during the lifetime of the LDL?

b. What assumptions did you make?

5-3 Red blood cells (RBC) carry carbon dioxide to be exhaled from the lungs as bicarbonate ions (anions). In the lung, each bicarbonate anion is exchanged for a chloride ion. Each RBC transports 10^9 bicarbonate ions across its plasma membrane in 50 milliseconds. This is accomplished by 1,200,000 transmembrane proteins in each RBC plasmalemma.

a. How many chloride ions are moved into an RBC per protein per second?

b. What assumptions did you make?

5-4 Transmembrane proteins contain alpha helices (usually made of nonpolar amino acids) that span the 4 nm lipid bilayer of a cell membrane. One peptide bond in the alpha helical part of the protein has a 1.5 angstrom rise and a 100 degree rotation.

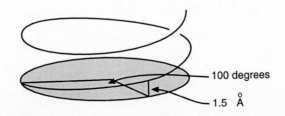

100 degrees

1.5 Å

a. How many amino acids are needed for one complete turn around the alpha helical spiral?

b. How much linear distance is traversed in one turn of the spiral?

c. What is the minimum number of turns needed to span the bilayer?

d. What is the minimum number of amino acids needed for the alpha-helix spanning component of these proteins?

e. What assumptions did you make?

5-5 A lipid bilayer has about 5 million lipid molecules (phosphatidyl choline) per square micrometer. The thickness of a lipid bilayer is 5 nm. All of the lipid molecules are formed in the cytoplasm and are inserted into the cytoplasmic side of the bilayer of the endomembrane system. In order to equalize the lipids in each bilayer, a protein (phospholipid translocator) called *flippase* moves lipid molecules from the cytoplasmic side to the noncytoplasmic (luminal) side. The Golgi, components of the endomembrane system, make spherical vesicles that have an external diameter of about 50 nm.

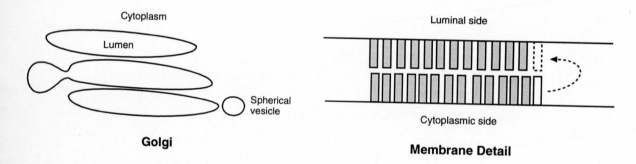

Golgi

Membrane Detail

a. What is the external surface area of this vesicle?

b. What is the internal surface area of this vesicle?

c. How many molecules make up the inner half of the lipid bilayer?

d. How many lipid molecules needed to be flipped to make this vesicle?

e. What assumptions did you make?

5-6 Animal cells contain calcium. Calcium ions free in the cytoplasm have a concentration of 10^{-7} M, but this can rise to as much as 5×10^{-6} M. Animal cells also contain *calmodulin*, a protein that binds calcium ions. A cell of 50 μm diameter can contain 10^7 molecules of calmodulin, each of which can bind four calcium ions. Assume that calmodulin is the means that the cell uses to lower the concentration of free calcium from 5×10^{-6} M back to 10^{-7} M.

   a. In the cell, what is the maximum number of calcium ions that can be bound?

   b. Can calmodulin bind all of the calcium ion difference found in the cell?

   c. What assumptions did you make?

5-7 In a cell with a diameter of 45 μm, a "resting" calcium ion concentration is 10^{-7} M. When the cell is activated, the internal stores of calcium can be released to the cytosol or calcium can enter through a transmembrane protein (pore). Suppose when activated the calcium ion concentration increases to 5×10^{-6} M, and later it returns to the resting concentration. One way the cell returns to its resting state is to pack Ca^{+2} onto a protein called *calsequestrin* inside a compartment (calcium-sequestering compartment). Each molecule of calsequestrin can bind 50 molecules of Ca^{+2}.

   a. What is the minimum number of calsequestrins needed to deactivate this cell?

   b. What assumptions did you make?

5-8 A fluorescing molecule emits light. A fluorescing molecule that binds at the cell membrane can be used to dye cells. In order to see the cell easily in a microscope field, a concentration of 10 molecules per square micrometer of cell membrane is needed.

   The *morula* is an early developmental stage of most animals. The morula has the same overall volume as the egg but has undergone synchronous divisions to yield a ball of 16 cells. The human egg can be modeled as a 100 μm diameter sphere.

   a. How many cell divisions occurred to produce the morula?

   b. Calculate the radius, surface area, and volume of both the egg and one morula cell.

   c. How many molecules of dye are needed for the egg?

   d. How many molecules of dye are needed for the morula cells?

   e. The ratio of the volume of the egg to the volume of one morula cell is 16. What is the ratio of the radius of the egg to the radius of the morula cell?

Chapter 5

f. What is the ratio of the surface area of the egg to the surface area of the morula cell?

g. What assumptions did you make?

5-9 Live cells have ions inside them that are at a concentration different from the cell's environment. Because of the charge separation the cell membrane has an electrical potential, with the inside being more negative compared to the environment side. Plant and animal cell membranes have a range of electrical potentials between -100 to -200 millivolts (mV). Because of this membrane potential, positive ions tend to diffuse into the cell and negative ions tend to diffuse out. To maintain life, ions are moved against this diffusional gradient. This moving of ions against a diffusional gradient is called *pumping* and requires biological energy.

Plant cell cytoplasm typically has a potassium ion concentration of 150 mM. The external environment has $[K^+] = 10$ mM. The plant cell can be modeled by a box 25 μm by 25 μm by 100 μm with 15% of the volume being cytoplasmic. The measured membrane potential is -150 mV; assume it remains constant. It costs 23,160 calories to move 1 mole of ions against a 1 V membrane potential.

a. What is the volume of the cell?

b. What is the volume of the cytoplasm?

c. How many moles of potassium exist in the cell's cytoplasm above what would be there at concentration 10 mM?

d. How many calories were required to move these potassium ions across the membrane?

e. What assumptions did you make?

5-10 *Liposomes* are laboratory-prepared artificial membranes. Liposomes can be made in a variety of sizes and can be made so that they have transmembrane proteins (pores). The contents of the liposomes can also be known. A lab makes liposomes that are 4 μm diameter spheres that have an average of ten protein pores. Each has internal potassium ion concentration 100 mM. Each protein pore will transport 10^7 potassium ions per second. The pores stay open an average of 0.3 second and stay closed an average of 2 seconds.

a. How many potassium ions are in a liposome initially?

b. How much time is required for the potassium ions in the liposome to come to equilibrium with their environment? (Assume the environment is relatively large and potassium-free.)

c. What assumptions did you make?

5-11 A rat mast cell contains hundreds of vesicles that contain histamine. A localized stimulus will cause vesicles to fuse with the cell membrane and dump their contents outside the cell. Mast cells and vesicles can be modeled as spheres with diameters 21 μm and 1.5 μm, respectively.

Assume a localized stimulus caused 50 vesicles to dump their contents.

a. What percentage temporary increase in mast cell surface occurred?

b. What assumptions did you make?

5-12 A sea urchin egg is 100 μm in diameter. Prior to fertilization, the pH of the egg is 6.8 and the cytosolic calcium ion concentration is 10^{-8} M. After fertilization, the pH increases to 7.25 and the calcium ions increase to 5×10^{-7} M. These changes occur in 200 seconds.

a. How many hydrogen ions were pumped out to achieve this pH?

b. How many calcium ions were imported to the cytosol?

c. What is the hydrogen ion export rate?

d. What is the calcium ion import rate?

e. What was the hydrogen ion net export rate per square micrometer of cell membrane?

f. What assumptions did you make?

5-13 *Glycophorin* is a single-pass transmembrane protein in red blood cells (RBC). The protein component of glycophorin is 131 amino acids long and binds carbohydrate on the outside (noncytoplasmic side) of the protein. The approximately 100 modified sugar residues that are attached near the end of each glycophorin account for 60% of the molecule's mass. The average molecular weight of an amino acid is 130.

a. What is the average molecular weight of each modified sugar residue on the glycophorin?

b. An RBC contains an average of 6×10^5 glycophorin molecules. How many modified sugar residues are found attached to glycophorins in one RBC?

c. How many grams does the protein component of glycophorin weigh in one RBC?

d. What assumptions did you make?

Chapter 5

CHAPTER 6

Respiration

Answers to Chapter 6 begin on page 208.

6-1 The respiration of glucose is described by the equation

$$C_6H_{12}O_6 + 6\,O_2 + 6\,H_2O \longrightarrow 6\,CO_2 + 12\,H_2O$$

An isotope of oxygen (^{18}O) can be used in the molecular oxygen as a tracer to determine the role of oxygen. When this is done, the left side of the equation above becomes

$$C_6H_{12}O_6 + 6\,{}^{18}O_2 + 6\,H_2O$$

What compound(s) gave rise to the

a. carbon in CO_2?

b. oxygen in CO_2?

c. oxygen in H_2O?

d. hydrogen in H_2O?

e. What percentage of the H_2O formed would contain ^{18}O?

f. What percentage of the oxygen in the CO_2 formed would contain ^{18}O?

g. What is the molecular weight of the CO_2 generated in the presence of $^{18}O_2$?

h. What is the molecular weight of the CO_2 generated in the presence of $H_2^{18}O$?

i. What is the molecular weight of the H_2O generated in the presence of $^{18}O_2$?

j. What assumptions did you make?

6-2 Several biopolymer monomers can be used as respiratory substrates. Complete the following equations for respiration:

SUGAR

Glucose: $C_6H_{12}O_6 +$ _____ $O_2 \longrightarrow$ _____ $CO_2 +$ _____ H_2O

FATTY ACID

Stearic acid: $C_{18}H_{36}O_2 +$ _____ $O_2 \longrightarrow$ _____ $CO_2 +$ _____ H_2O

AMINO ACID

Leucine: $C_6H_{13}O_2N +$ _____ $O_2 \longrightarrow$ _____ $CO_2 +$ _____ $H_2O +$ _____ NH_3

The ratio of CO_2 produced per O_2 used is called the *respiratory quotient* (RQ).

a. Calculate the respiratory quotient for each of the respiratory substrates.

b. What assumptions did you make?

6-3 You eat a candy bar that has 180 calories. This energy is converted during respiration to ATP. The reaction

$$ADP + P_i \longrightarrow ATP$$

requires 7.3 kcal/mol.

One "dietary" calorie is equal to 1000 "chemical" calories. Respiration is maximally about 39% efficient in converting substrate calories to ATP calories.

a. How many ATPs could your body make from this candy bar?

b. What assumptions did you make?

6-4 The first two stages in respiration are glycolysis and Kreb's cycle. For each molecule of glucose input,

- 10 NAD^+ molecules are reduced

- 2 FADs are reduced

- 4 ADPs are phosphorylated.

The free energy $\Delta G^{o'}$ of the relevant "respiratory reactions" is:

Reaction	$\Delta G^{o'}$ (kcal/mol)
$NADH + H^+ + \frac{1}{2} O_2 \longrightarrow NAD^+ + H_2O$	− 51.7
$FADH_2 + \frac{1}{2} O_2 \longrightarrow FAD + H_2O$	− 43.4
$ATP \longrightarrow ADP + P_i$	− 7.3
$Glucose + 6 O_2 \longrightarrow 6 CO_2 + 6 H_2O$	−686

a. How much of the energy from one mole of glucose is conserved in ATP during the first two stages?

b. How much of the glucose energy is conserved in the reduced coenzymes NAD^+ and FAD?

The third stage of respiration is oxidative phosphorylation.

c. A total of 32 ATPs per glucose molecule are made during oxidative phosphorylation from the reduced coenzymes. How much of the glucose energy is conserved in ATP at the end of all three stages?

d. What percentage of the total ATP energy is converted by the oxidative phosphorylation of the reduced coenzymes?

e. What percentage of the glucose energy was lost as heat?

f. What assumptions did you make?

6-5 For an average biology student who weighs 70 kilograms, each liter of oxygen consumed releases 4.8 usable kilocalories of energy. Each mole of glucose consumed yields 686 kcal of energy.

During an awake but restful period, the biology student uses 20 kilocalories per kilogram of body weight per day (basal metabolic rate). During a normally active period, the metabolic rate averages 40 kcal/kg body wt/day for a male and 35 kcal/kg body wt/day for a female. During sleep, the metabolic rate averages 1.2 kcal/kg body wt/day.

a. How many liters of O_2 is consumed per hour by a student who is resting but awake?

b. If air is 20% oxygen, what is the minimum amount of air moved per hour by a student when resting?

c. If a student were respiring only glucose, how many grams of glucose would be needed per hour when at the basal rate?

d. A female biology student, who is a distance runner, uses 14.3 kilocalories per kg of body weight per hour when running. If she practices her running techniques for 2 hours a day in preparation for a race, how many kilocalories would she use per day?

e. What assumptions did you make?

Chapter 6

CHAPTER 7

Photosynthesis

Answers to Chapter 7 begin on page 211.

7-1 Chloroplasts use the pH difference across the thylakoid membranes (especially the stroma lamellae) to drive ATP synthesis (the phosphorylation of ADP). Three hydrogen ions are moved through an ATP-generating enzyme to convert one ADP and one P_i to one ATP. Three ATPs are needed to fix one CO_2 to a carbohydrate carbon.

A measured photosynthetic rate in *Atriplex* (a weedy plant) is 28 micromoles of CO_2 fixed per square meter of soil surface per second.

a. How many CO_2 molecules are being fixed by *Atriplex* per square meter per second?

b. How many ATP molecules are required for this CO_2 fixation rate?

c. How many hydrogen ions are moved across the thylakoid membranes in a square meter of soil surface to yield the *Atriplex* photosynthetic rate?

d. What assumptions did you make?

7-2 The fixing of carbon in photosynthesis varies in regard to the amount of ATP needed for each carbon fixed.

• C_3 plants use 3 ATPs per carbon.

• C_4 plants use 5 ATPs per carbon.

• CAM plants use 5.5 ATPs per carbon.

If ATP yields 7.3 kilocalories per mole, how many ATP calories are needed to create 1 mole of glucose (686 kcal/mol) by

a. a C_3 plant?

b. a C_4 plant?

c. a CAM plant?

d. Which type plant uses the most ATP energy in the making of glucose?

e. What assumptions did you make?

7-3 Photorespiration can reduce the rate of photosynthesis by 50%. Furthermore, photorespiration is energetically quite costly since NADPH and ATP are utilized without carbon fixation.

Chapter 7

The energy costs of photosynthesis for each CO_2 fixed by a

- C_3 plant are 3 ATPs and 2 NADPHs

- C_4 plant are 5 ATPs and 2 NADPHs.

The energy costs of photorespiration for each O_2 fixed by a

- C_3 plant are 2 ATPs and 2.5 NADPHs.

When in air, plants that photorespire typically have a 3:1 ratio of carboxylation to oxidation (photosynthesis to photorespiration).
The free energy of the two relevant reactions is

$$NADPH \longrightarrow NADP^+ \qquad \Delta G = -51.7 \text{ kcal/mol}$$

$$ATP \longrightarrow ADP + P_i \qquad \Delta G = -7.3 \text{ kcal/mol}$$

Compare the energy costs of making a mole of glucose in a C_3 and a C_4 plant.

a. How many kilocalories are needed by a C_3 plant that does not photorespire?

b. How many kilocalories are needed by a C_4 plant?

c. How many kilocalories are needed by a C_3 plant that photorespires?

d. What assumptions did you make?

7-4   When light is shined on a leaf, it causes hydrogen ions to be pumped into discs called *thylakoid lumens*. The ions then diffuse out through a protein and in the process an ATP molecule is made for every three hydrogen ions. While illuminated, inside the disc the pH can be as low as 4. Outside the disc the pH is about 7.2.
A thylakoid lumen can be modeled as a short cylindrical disk or rod 80 Å long by 5000 Å in diameter.

a. How many hydrogen ions are found in one thylakoid lumen of this size at pH 4?

b. How many are found at pH 7.2?

c. How many more ATP molecules can be made from the disc described above after the light is turned off?

d. What assumptions did you make?

7-5 Light is important in biology for photosynthesis. There are two different ways that light is described in physics.

In the first description, light travels in waves at a fixed speed $c = 2.998 \times 10^8$ meters per second. The *wavelength* is the distance from peak to peak of a light wave, and corresponds to the color of the light. The wavelength is given by λ (the Greek letter *lambda*). The wavelength varies from 400 nm to 700 nm for light in the visible range, with blue light having $\lambda = 450$ nm and red light having $\lambda = 680$ nm.

The *frequency* is given by ν (the Greek letter *nu*). The frequency is the number of peaks that pass a point in a given time. Frequency is related to wavelength by the formula

$$\nu = \frac{c}{\lambda}$$

In the second description, light travels in particles called *photons* or *quanta*. Using this description it makes sense to speak of a mole of light as 6.02×10^{23} photons.

The energy of one photon of light is given by

$$E = \frac{\hbar c}{\lambda} = \hbar \nu$$

where $\hbar$ is a conversion factor called *Planck's constant*; $\hbar = 1.583 \times 10^{-34}$ calorie seconds.

In the laboratory, light with a very narrow wavelength range can be used for experiments. An actinic light (activating light) can be made to flash for only 1 picosecond.

a. What is the frequency of 450 nm light?

b. What is the frequency of 680 nm light?

c. What is the energy of a mole of 450 nm light?

d. What is the energy of a mole of 680 nm light?

e. How many peaks of 450 nm light pass a point in 1 picosecond?

f. How many peaks of 680 nm light pass a point in 1 picosecond?

g. What assumptions did you make?

7-6 In the laboratory, red light can be given to a photosynthetic plant. This red light has an energy value of 41 kcal/mol. Assume one entire mole of red light did chemistry (no quanta were lost).

During photosynthesis, two processes move hydrogen ions into the thylakoid lumens:

Noncyclic electron transport

$$2\,H_2O + 2\,NADP^+ + 4\,H^+_{stroma} + 8\,quanta \longrightarrow 2\,NADPH + 8\,H^+_{thylakoid} + O_2$$

Cyclic electron transport

$$2\,H^+_{stroma} + 2\,quanta \longrightarrow 2\,H^+_{thylakoid}$$

As hydrogen ions diffuse back out of the lumens, ATP molecules are produced:

Chemiosmosis

$$9\,H^+_{thylakoid} + 3\,ADP + 3\,P_i \longrightarrow 9\,H^+_{stroma} + 3\,ATP$$

These ATP molecules, along with the NADPH products of noncyclic electron transport, convert CO_2 to glucose:

Calvin cycle

$$6\,CO_2 + 18\,ATP + 12\,NADPH \longrightarrow Glucose + 18\,ADP + 18\,P_i + 12\,NADP^+$$

a. In order to produce a molecule of glucose, what is the theoretical minimum number of quanta required, and how are they divided between noncyclic and cyclic electron transport?

b. How many moles of glucose are produced from this one mole of red light?

c. How many moles of O_2 are evolved?

d. How many moles of CO_2 are fixed?

e. How would any of the above numbers change if a mole of blue light (72 kcal/mol) were used instead of the red light?

f. What assumptions did you make?

7-7   Photosynthesis can be studied in the laboratory. The fixation of carbon can be measured using $^{14}CO_2$, which allows us to determine that ten photons are needed to fix one molecule of CO_2 into the sugar glucose. Glucose weighs 180 grams per mole and yields energy at the rate of 686 kcal/mol.

Light itself has energy, the amount depending on the color. The energy of red light is 41 kcal/mol. The energy of blue light is 72 kcal/mol.

Assume all the photons in one mole of light were utilized to do chemistry (none were lost).

a. How many grams of CO_2 were fixed?

b. How much glucose was made?

c. What is the efficiency of conversion of red radiant energy to chemical carbohydrate energy in the glucose?

d. What is the efficiency of conversion of blue radiant energy to carbohydrate energy?

e. What assumptions did you make?

7-8 The maximum photosynthetic rates for plants in natural conditions are:

Type of Plant	Species	Rate
CAM	Century plant	2.4
C_3	Soybean	20
C_4	Corn	40

The rate is measured in micromoles of CO_2 per square meter per second.

Sunlight, on a clear summer day, can supply as many as 2300 micromoles of photons per square meter per second.

In an experimental setting, at least 12 photons are required to fix one CO_2.

a. If the photosynthetic machinery were 100% efficient, how many moles of CO_2 could be fixed per square meter per second?

b. What is the efficiency of the three types of plants (CAM, C_3, and C_4) in converting radiant energy of photons into chemical energy of glucose?

c. What assumptions did you make?

7-9 Light is important in biology for photosynthesis. There are two different ways that light is described in physics.

In the first description, light travels in waves at a fixed speed $c = 2.998 \times 10^8$ meters per second. The *wavelength* is the distance from peak to peak of a light wave, and corresponds to the color of the light. The wavelength is given by λ (the Greek letter *lambda*). The wavelength varies from 400 nm to 700 nm for light in the visible range, with blue light having $\lambda = 450$ nm and red light having $\lambda = 680$ nm.

The *frequency* is given by ν (the Greek letter *nu*). The frequency is the number of peaks that pass a point in a given time. Frequency is related to wavelength by the formula

$$\nu = \frac{c}{\lambda}$$

Chapter 7

In the second description, light travels in particles called *photons* or *quanta*. Using this description it makes sense to speak of a mole of light as 6.02×10^{23} photons.

The energy of one photon of light is given by

$$E = \frac{\hbar c}{\lambda} = \hbar \nu$$

where $\hbar$ is a conversion factor called *Planck's constant*; $\hbar = 1.583 \times 10^{-34}$ calorie seconds.

The maximal amount of light energy received in the midwestern United States is 2300 micromoles of photons per square meter per second, of which about 90% is absorbed. Assume the average wavelength of this light is 550 nm.

This radiant light energy can cause the temperature in the leaf of a plant to increase. In crop plants, about 1.25 grams of water exist in each 100 cm^2 of leaf surface area. The heat of vaporization of water is 540 calories per gram. The definition of a calorie is the amount of heat required to raise the temperature of a gram of water from $15°$ to $16°C$. Assume the outside temperature is $25°C$.

a. How much time is required to start to boil a leaf's water?

b. How much time is required to boil off all of the water in a leaf?

c. What assumptions did you make?

7-10   Light is important in biology for photosynthesis. There are two different ways that light is described in physics.

In the first description, light travels in waves at a fixed speed $c = 2.998 \times 10^8$ meters per second. The *wavelength* is the distance from peak to peak of a light wave, and corresponds to the color of the light. The wavelength is given by λ (the Greek letter *lambda*). The wavelength varies from 400 nm to 700 nm for light in the visible range, with blue light having $\lambda = 450$ nm and red light having $\lambda = 680$ nm.

The *frequency* is given by ν (the Greek letter *nu*). The frequency is the number of peaks that pass a point in a given time. Frequency is related to wavelength by the formula

$$\nu = \frac{c}{\lambda}$$

In the second description, light travels in particles called *photons* or *quanta*. Using this description it makes sense to speak of a mole of light as 6.02×10^{23} photons.

The energy of one photon of light is given by

$$E = \frac{\hbar c}{\lambda} = \hbar \nu$$

where $\hbar$ is a conversion factor called *Planck's constant*; $\hbar = 1.583 \times 10^{-34}$ calorie seconds.
   In the laboratory, light with a very narrow wavelength range can be used for experiments. One mole of an actinic light (activating light) that has a wavelength of 680 nm was used to excite chlorophyll, and caused fluorescence measured at a wavelength of 690 nm. The chlorophyll was isolated, and therefore could do no photochemistry.

a. What is the amount of energy (in kilocalories) in one mole of actinic red light?

b. What is the amount of energy in the light that was fluoresced (assuming maximal fluorescence)?

c. What is the amount of energy lost as heat?

d. What percentage of the red light energy was lost as heat?
   A photon of blue light will energize an electron from chlorophyll to a level comparable to a photon of red light. Suppose blue light energy also caused fluorescence measured at a wavelength of 690 nm.

e. What percentage of the blue light energy was lost as heat (again assuming maximal fluorescence)?

f. What assumptions did you make?

Mitosis

Answers to Chapter 8 begin on page 216.

8-1   A yeast haploid genome has about 12.5 million base pairs (bp) distributed among 15 chromosomes. The chromosomes are attached to microtubules by 220 bp when the cell is undergoing mitosis. The chromosomes are condensed by attachment to histones in nucleosomes. Each nucleosome binds 140 bp. There are 60 bp in each DNA link between adjacent nucleosomes.

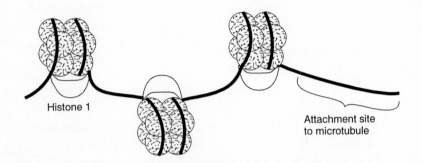

Nucleosomes

a. How many nucleosomes are required to fully condense the chromosomes in a cell undergoing mitosis?

b. What assumptions did you make?

8-2   In the human brain, an estimated average rate of 250,000 nerve cells are produced each minute during gestation and through 6 months after birth.

a. How many nerve cells are in a 6-month-old human brain?

b. What assumptions did you make?

8-3   The DNA in one human cell nucleus has 6 billion base pairs with an S phase lasting about 9 hours. One DNA polymerase can catalyze at the rate of 100 nucleotides added per second.

a. How many copies of DNA polymerase are required to complete the S phase under the conditions listed above?

b. What assumptions did you make?

Chapter 8

8-4 One of the earliest stages of fertilization that can be measured in a sea urchin egg is a depolarization of the membrane around the egg. This occurs in 2 seconds. Egg and sperm nuclei fuse at around 1200 seconds and the first division of the zygote is at 5500 seconds.

    a. How many times longer does the process of nuclear fusion take than membrane depolarization?

    b. What is the maximum amount of time needed to synthesize a complete set of chromosomes in the sea urchin zygote prior to its first division?

    c. What assumptions did you make?

8-5 The DNA that is found in the nucleus of each of your cells has 6 billion nitrogen base pairs, distributed among 46 chromosomes. In some cells, this DNA is being replicated while you read this problem. Each nitrogen base added to a new strand uses two ATP equivalents. You have a net gain of 36 ATPs for each glucose you respire.

    a. How many glucose molecules are used to convert one of your cells from G_1 to G_2?

    b. How many glucose molecules are used to duplicate one chromosome?

    c. What assumptions did you make?

8-6 Mammalian cells have a nucleus (a 7 μm diameter sphere) that contains on average 11 pores per square micrometer of nuclear membrane. During the DNA synthesis phase of the cell cycle, about 1 million histone proteins must be imported through these pores every 3 minutes.

    a. What is the surface area of the nuclear membrane?

    b. How many nuclear pores are in the nuclear membrane?

    c. What is the rate of histone import per pore?

    d. What assumptions did you make?

8-7 For some kinds of experiments, biologists use isolated cells grown in culture. Cells differ significantly in their cell doubling times (one cell dividing into two).

Plant cells can double every	18 hr
Animal cells (matrix requiring) can double every	18 hr
Animal cells (nonmatrix requiring) can double every	14 hr
Yeast cells can double every	2 hr
Bacteria cells can double every	20 min

a. Starting with one cell of each type above, how many cells of each type would you have after one week?

b. What assumptions did you make?

8-8   Here is data for the rate of cell divisions of a *Xenopus laevis* (an African frog) fertilized egg. The numbers in the first column give total elapsed time in hours from when the egg was fertilized.

Time in Hours	Number of Cells	Remarks
0	1	egg is fertilized
1/2	2	first cleavage
4	64	polarization of dorsal/ventral orientation
6	10,000	blastula
10	30,000	gastrula
19	80,000	neurula
32	170,000	somite formation
110	10^6	feeding tadpole

The first 12 cycles of cell division are synchronous.

a. How many cell divisions took place in the first 4 hours?

b. How many cells exist after 12 cycles?

c. Calculate the rate of cell divison during each of the seven time intervals (change in number of cells per change in time).

d. Between which of the stages is the rate of cell division the most rapid?

e. What assumptions did you make?

8-9 In a textbook, a micrograph of a fertilized hamster egg showed a spherical cell. A note indicated it was magnified one thousand times (1000 ×). Measuring with a ruler revealed a diameter of 4 mm.

    a. What is the diameter of the hamster zygote?

    b. What is the volume of the cell?

       After several synchronous divisions, the zygote becomes a morula of 16 equal-volume cells with no total volume gain or loss.

    c. What is the volume of one morula cell?

    d. What is the diameter of one morula cell?

    e. What assumptions did you make?

8-10 *Microtubules* play a role in the migration of chromosomes to opposite ends of a mitosing cell during anaphase. Microtubules are hollow tubes 24 to 25 nm in diameter composed of 13 parallel rows. The parallel rows are called *protofilaments* and are made of heterodimers called *tubulin subunits*. A tubulin heterodimer is 8 nm in length.

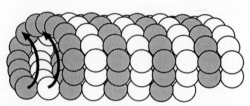

Microtubule

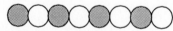

Tubulin protofilament

Tubulin heterodimer subunit

    According to a current hypothesis, microtubules are disassembled at the centromere by the removal of these subunits. This shortens the microtubules, drawing the chromosomes along.

The human somatic cell has 46 chromosomes, and 15 to 35 microtubules are attached to each one. Anaphase usually takes from 2 to 10 minutes and moves each chromosome about 2 μm.

a. How many tubulin subunits are disassembled from microtubules during anaphase?

b. What is the rate of tubulin dissolution in a human cell undergoing mitosis?

c. What assumptions did you make?

8-11 The haploid human genome is estimated to contain 3 billion base pairs. These bases are known to be contained in 23 chromosomes. Assume a 50:50 split between AT pairs and GC pairs on an "average" chromosome.

a. How many H bonds bind the average chromosome?

A covalent bond is 110 times the strength of an individual H bond.

b. How many covalent bonds would be required to yield the same bond energy value as is in one chromsome?

c. What assumptions did you make?

8-12 In a mature spinach leaf

• one cell contains an average of 171 chloroplasts,

• each chloroplast contains an average of 32 genome copies,

• the genome of the spinach chloroplast contains 150 kbp (kilo base pairs),

• a reasonable number of mitochondria in each leaf cell is 10,

• the number of genomic copies in each mitochondrian is about 10,

• plant mitochondria have genomic sizes from 200 to 2400 kbp (assume it is 1000 kbp).

One measurement suggests a mature spinach leaf cell contains 3.6 billion base pairs.

a. What percentage of the total base pairs are found in the chloroplasts?

b. What percentage of the total base pairs are found in the mitochondria?

c. What percentage of the total base pairs are found in the nucleus?

d. What assumptions did you make?

CHAPTER 9

Mendelian Genetics

Answers to Chapter 9 begin on page 220.

9-1   As a woman ages, the probability of Down syndrome in her offspring increases.

Age of Woman	Down Incidence
20	1/2,300
25	1/1,600
30	1/1,200
37	1/290
43	1/100
46	1/46

a. Plot the data.

b. Does the incidence of Down syndrome appear to increase linearly or exponentially with age?

c. How much more likely is it that a woman aged 46 will bear a child with Down syndrome than a woman 26 years younger at age 20?

d. What assumptions did you make?

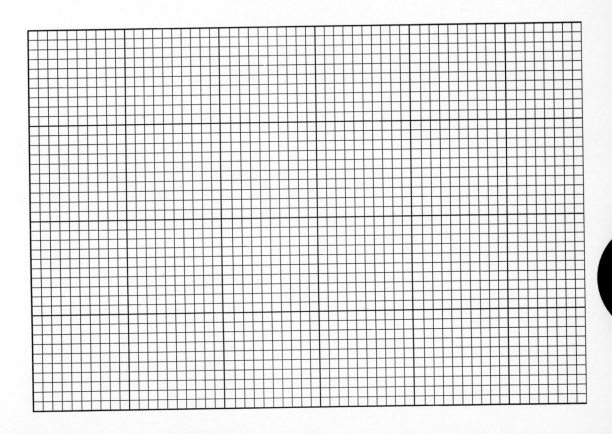

9-2   The Punnett square is a typical method to illustrate the possible genotypes of a cross when a small number of genes are involved. The use of a Punnett square becomes unwieldy when a larger number of genes are involved. Suppose you were to use a Punnett square to illustrate Mendel's pea experiments with seven genes segregating independently.

   a. How many different gametes from a heterozygous parent are possible?

   b. How many little boxes would be required in the square?

   c. How many different genotypes could be produced?

   d. How many different phenotypes could be produced?

   e. What assumptions did you make?

9-3   In a monohybrid cross, several ratios of progeny phenotypes can occur. What are the phenotypic ratios of the following crosses? Assume A is dominant to a.

   a. $AA \times aa$?

   b. $AA \times AA$?

   c. $aa \times aa$?

   d. $Aa \times aa$?

   e. $AA \times Aa$?

   f. $Aa \times Aa$?

      If homozygous dominant is lethal, what is the phenotypic ratio of

   g. $Aa \times Aa$?

      If homozygous recessive is lethal, what is the phenotypic ratio of

   h. $Aa \times Aa$?

      If heterozygotes are intermediate (intermediate inheritance is evidenced), what is the phenotypic ratio of

   i. $Aa \times Aa$?

   j. What assumptions did you make?

9-4 One hypothesis is that human skin color is governed by three equally important genes, each of which has two alleles that do not display dominance. This is an example of polygenic inheritance—a common situation when a trait shows a continuous gradation in phenotypes.

a. How many different genotypes are possible?

b. How many different phenotypes are possible?

c. What assumptions did you make?

9-5 Coat color in rabbits is determined by a gene with at least four alleles (multiple allelism) that display a hierarchy of dominance. The known alleles are

C	full color	most dominant
C^{ch}	chinchilla	
C^h	himalayan	
c	albino	most recessive

What is the number of different genotypes possible for a

a. full-color rabbit?

b. chinchilla rabbit?

c. himalayan rabbit?

d. albino rabbit?

e. A particular mated pair consists of a full-color rabbit and a himalayan rabbit. Their first litter contained 4 rabbits, 3 full-color and 1 chinchilla. They continued to have many more litters. What would be the phenotype(s) and frequency(ies) expected?

f. What assumptions did you make?

9-6 Mendel's dwarf peas might have been caused by a leaky mutant of the Le gene. Short or dwarf peas are caused by a recessive le allele that allows a little of the plant hormone GA_1 to be made. Le is dominant and lets a larger amount of GA_1 be made, so that the plants grow taller.

Another gene, Na, allows the synthesis of GA_1, but its allele does not allow the production of GA_1. A plant with the homozygous recessive $nana$ is extremely short and phenotypically different from the dwarf phenotype. So while Mendel worked with one gene, clearly at least two different genes determine height due to GA_1 concentrations.

Chapter 9

a. If a plant with genotype *NanaLele* is self-pollinated, what phenotypes can occur and in what ratio would we find their frequencies?

b. What assumptions did you make?

9-7   In corn (*Zea mays*), 20 genes are known to be involved in formation of anthocyanin pigment in kernels. Two of them, *P* and *C*, are involved as follows: The dominant allele *P* yields purple color and the recessive allele *p* yields bronze.

The gene *C* allows pigment formation, but its allele *c* blocks it. A plant with the homozygous recessive *cc* cannot form anthocyanins and the phenotype is brown.

a. If a doubly heterozygous plant is self-pollinated, what genotypes can occur and in what ratio would we find their frequencies?

b. What phenotypes can occur and in what ratio would we find their frequencies?

c. What assumuptions did you make?

9-8   William syndrome occurs when a deletion is evidenced in one of the two #7 chromosomes in humans. (A deletion in both chromosomes is lethal.) Affected individuals have good musical and verbal ability but lack normal spatial and mathematical abilities and read very poorly. The deletion leads to an IQ in the range 40–100. About one person in 20,000 has William syndrome.

a. Is it possible to be a carrier of William syndrome?

If two William syndrome people wish to become parents, what is the probability of

b. a normal child?

c. a child with William syndrome?

d. a spontaneous abortion?

e. What assumptions did you make?

9-9   In corn (*Zea mays*), the color of the kernel is due to two units of genetic information in the developing ear and one unit from the pollen, developed in the tassel. The two units in the ear result from a haploid nucleus that underwent mitosis. One gene for kernel color is *P* (purple) or *p* (white), which demonstrates dominance/recessiveness.

   a. If a heterozygous corn plant is self-pollinated, what are the phenotypes of kernel color and their frequencies?

   b. If a heterozygous ear-producing plant is cross-pollinated with a homozygous *p* pollen producer, what are the phenotypes of kernel color and their frequencies?

   c. If a homogyzous *p* plant is the ear producer that is crossed with a heterozygous plant as the pollen producer, what are the phenotypes of kernel color and their frequencies?

   d. If kernel color were due to a gene that demonstrates intermediate inheritance and displays the colors *PPP*—dark purple, *PPp*—violet, *Ppp*—red, and *ppp*—white, what would be the phenotypes of kernel color and their frequencies in each of cases (a), (b), and (c)?

   e. What assumptions did you make?

9-10 In wild-type fruit flies, gray body *G* and regular wings *R* are dominant to black body *g* and vestigial wings *r*.
   A wild-type fruit fly homozygous for gray body and regular wings is crossed with a black bodied vestigial winged fly.

   a. Show the cross.

   b. If the resulting heterozygous offspring were crossed with a known homozygous recessive individual and the two traits were not linked, what would be the phenotype(s) and the frequency(ies) of the phenotypes of this cross?

   However, the above cross (heterozygote with a known homozygous recessive) yielded the following phenotypes and frequencies, independent of sex:

   40% of the offspring were gray-regular

   40% of the offspring were black-vestigial

   10% of the offspring were gray-vestigial

   10% of the offspring were black-regular

   c. What kind of inheritance is demonstrated?

   d. How many map units is body color from wing length?

e. If a third trait (eye color) is known to be 10 map units from wing length and 10 map units from body color, show the sequence of genes on this chromosome.

f. What assumptions did you make?

9-11 Tomatoes have 12 chromosomes. Chromosome 1 has four genes that alter fruit attributes. Each of them shows dominant/recessive characters. You have a seed stock that is known to be quadruply heterozygous for these genes. You plant a field of tomatoes (which are cross-pollinated) in order to sell these novelty tomatoes at a roadside stand.

a. How many different gametes can be produced?

b. How many different genotypes can be produced?

c. How many different phenotypes can be produced?

   The distance between gene A and gene B is 17 centimorgans, between gene B and gene C is 31 centimorgans, and between gene C and gene D is 14 centimorgans. Genes B and D appear to assort independently.

d. What is the sequence of genes on chromosome 1?

e. What is the distance between gene A and gene C?

f. What assumptions did you make?

9-12 Crossovers appear in humans at a higher frequency in women than in men. Below is the map-unit distance estimated between some genes on chromosome 12 in women and in men

| Women: | A | 22 | B | 26 | C | 29 | DE | 42 | F | 10 | G |
| Men: | A | 9 | B | 12 | C | 14 | DE | 20 | F | 23 | G |

a. What percentage more likely is a crossover in a woman than in a man?

b. If a map unit is about 1000 kilo base pairs, how many additional kilo base pairs exist in a woman's chromosome 12?

c. What assumptions did you make?

Population Genetics

Answers to Chapter 10 begin on page 225.

10-1 The χ^2 (chi-square) test from statistics is useful for testing hypotheses we might have about experimental data. Here are Mendel's data giving the results of F_1 crosses for seven characters in pea plants:

Character	Dominant Trait	Recessive Trait	Ratio
Flower color	purple	white	705:224
Flower position	axial	terminal	651:207
Seed color	yellow	green	6022:2001
Seed shape	round	wrinkled	5474:1850
Pod shape	inflated	constricted	882:299
Pod color	green	yellow	428:152
Stem length	tall	dwarf	787:277

In each case, we make the hypothesis that the ratio of dominant to recessive is 3:1. If this were exactly true for the $705 + 224 = 929$ flowers, we would expect $(3/4)929 = 696.75$ purple and $(1/4)929 = 232.25$ white, rather than 705 and 224. A more precise statement of the hypothesis is that the difference of the observed values from the expected values is due to random variation rather than to anything that is statistically significant.

To test the hypothesis for flower color, we first fill in the observed and expected values in the following chart:

	Dominant	Recessive
Observed (O)	705	224
Expected (E)	696.75	232.25

We compute χ^2 by finding for each type, dominant and recessive, $(O - E)^2/E$ and summing the results:

$$\chi^2 = \sum \frac{(O - E)^2}{E}$$

a. Compute the χ^2 value for flower color.

The closer the observed values are to the expected values, the smaller χ^2 is and the more likely the hypothesis is to be true.

Next we determine the number of degrees of freedom. This is one less than the number of types; in this case it is 1 since there are only two types: dominant and recessive.

The χ^2 table (see the Appendix F) gives, for different numbers of degrees of freedom, critical values for χ^2. In this case for 1 degree of freedom, only 5% of the time will the computed value of χ^2 be larger than 3.841. If the number computed in (a) is larger than this, we reject the hypothesis that the ratio of dominant to recessive flowers is 3:1.

b. Would we reject the hypothesis?

c. Choose one other character and test the same hypothesis with the χ^2 test.

d. What assumptions did you make?

10-2 The χ^2 (chi-square) test from statistics is useful for detecting linkage of genes during meiosis. If test crosses are made of *AABB* with *aabb*, the offspring of the first generation are all *AaBb*. If these offspring are then crossed with individuals of type *aabb*, their offspring have types *AaBb, aabb, Aabb,* and *aaBb*. Let us assume that the genes are equally viable. If there is no crossing over, we would expect to see offspring of each of these four types in the ratio 1:1:1:1.

Suppose in a particular experiment that 1000 such crosses are made, and the number of each type observed is listed as O in the table below. Let us test the hypothesis that they are close enough to a random sample from a population that is in the ratio 1:1:1:1.

	$AaBb$	$aabb$	$Aabb$	$aaBb$
O	319	174	191	316
E				

a. Enter the number that would be expected in each case in the row marked E.

b. Compute χ^2 by finding for each type $(O - E)^2/E$ and summing up the results:

$$\chi^2 = \sum \frac{(O - E)^2}{E}$$

The closer the observed values are to the expected values, the smaller χ^2 is and the more likely the hypothesis is to be true.

Next determine the number of degrees of freedom. This is one less than the number of types; in this case it is 3.

The χ^2 table (see the Appendix F) gives, for different numbers of degrees of freedom, critical values for χ^2. In this case for 3 degrees of freedom, only 5% of the time will the computed value of χ^2 be larger than 7.815. If the number computed in (b) is larger than this, we reject the hypothesis that the population is in the ratio 1:1:1:1. We would be wrong only 5% of the time.

c. Would we reject the hypothesis that the population is in the ratio 1:1:1:1?

d. Test the same data again against the hypothesis that the population is in the ratio 2:1:1:2.

e. Why would we cross hybrid offspring with the recessive type *aabb* rather than *AABB*?

f. What assumptions did you make?

10-3 Punnett and a co-worker studied genetics in sweet peas. They found one gene that controlled purple (P) or red (p) flowers and another that controlled long (L) or round (l) pollen. Parental true-breeding purple-flowered long-pollen plants were crossed with true-breeding red-flowered round-pollen plants. All the progeny (F_1) were purple-flowered long-pollen phenotypes. When the F_1 progeny were self-pollinated, the results were 284 purple long, 21 purple round, 21 red long, and 55 red round.

From these data it appears that the two genes are linked, or perhaps that the alleles are not equally viable. The χ^2 (chi-square) test from statistics is useful for testing hypotheses we might have about experimental data. Let us test first the hypothesis that the alleles are equally viable and that they segregated independently. We list the observed (O) and expected (E) numbers of each type in a table:

	PL	Pl	pL	pl
O	284	21	21	55
E				

a. What is the total number of progeny counted?

b. In what ratio would these progeny be if we assume the hypothesis?

c. Fill in row E with the number of each type that would be expected assuming the hypothesis.

d. Compute χ^2 by finding for each type $(O - E)^2/E$ and summing up the results:

$$\chi^2 = \sum \frac{(O - E)^2}{E}$$

The closer the observed values are to the expected values, the smaller χ^2 is and the more likely the hypothesis is to be true.

Next determine the number of degrees of freedom. This is one less than the number of types; in this case it is 3.

The χ^2 table (see the Appendix F) gives, for different numbers of degrees of freedom, critical values for χ^2. In this case for 3 degrees of freedom, only 5% of the time will the computed value of χ^2 be larger than 7.815. If the number computed in (d) is larger than this, we reject the hypothesis. We would be wrong only 5% of the time.

e. Based on the value you computed for χ^2, would you reject the hypothesis?

If the genes are linked, the purple round and red long are recombinant.

f. What fraction of the total number of progeny are recombinant?

Let us formulate another hypothesis, that the alleles are equally viable; that half the recombinants are purple round and the other half are red long; and that half the nonrecombinants are purple long and the other half red round.

g. Fill in row E with the number of each type that would be expected assuming this hypothesis.

h. Complete the χ^2 test. Would we reject this hypothesis?

i. What assumptions did you make?

10-4 Use the Hardy-Weinberg theorem for populations in equilibrium to answer the following questions.

Suppose a gene has two alleles, B and b, with B dominant. If the frequency of the allele B is 0.7,

a. What percentage of the population display the phenotype B?

b. What percentage of the population are carriers for the recessive allele b?

c. What percentage of the population display the recessive phenotype?

If 16% of people are left-handed and you randomly select a right-handed person (right-handedness is dominant),

d. What is the chance this individual is homozygous?

e. What is the chance this individual is heterozygous?

If you know you are a carrier of a recessive trait that is expressed in only 1% of the population,

f. What is the probability that your classmate is also a carrier?

g. What percentage of the population are not carriers?

A breeder of exotic tulips finds two tulips out of 200,000 on a 2-acre field that show a unique flower shape.

h. How many of the rest of the flowers are not true-breeding?

Suppose 68% of the population displays a dominant trait.

i. What is the frequency of the dominant allele?

j. What assumptions did you make?

10-5 In peas, the flower color purple (P) is dominant over the recessive color white (p). You perform three experiments.

In the first, you begin with an initial population in which all of the individuals are heterozygous and have purple flowers. You allow them to interbreed.

a. What is the distribution of PP, Pp, and pp in the first generation?

b. What is the distribution in the second generation?

The second experiment is the same as the first except all the purple-flowered peas are homozygous.

c. What is the distribution of PP, Pp, and pp in the first generation?

d. What is the distribution in the second generation?

For the third experiment, you begin with an initial population that is 60% PP and 40% Pp.

e. What is the distribution of PP, Pp, and pp in the first generation?

f. What is the distribution in the second generation?

g. What assumptions did you make?

10-6 Malic dehydrogenase can exist in two types, A and B. Fruit flies captured in Austin, Texas, have 95.7% type A alleles and 4.3% type B alleles. Xanthine oxidase in the same fruit flies has five types of alleles. Using the Hardy-Weinberg equation,

a. What percentage of the captured fruit flies demonstate heterozygosity?

b. What percentage demonstrate type B?

c. How many genotypes exist for xanthine oxidase in this population?

d. What assumptions did you make?

CHAPTER 11

Molecular Genetics

Answers to Chapter 11 begin on page 229.

11-1 The DNA in human nongametic cells contains 6 billion base pairs. It is estimated that about 10,000 DNA changes occur in each cell in one day. These are quickly repaired so that only a few (1 to 5) mutations accumulate in one cell in a year.

a. What percentage of the base pairs are altered each day?

b. What percentage of the DNA changes that occur in one cell in one year escape the proofreading and repair process?

c. What assumptions did you make?

11-2 Chemicals in the environment can cause mutations. A rather low level of mutation is caused by 1,2-epoxybutane. The rate is 0.006 mutants generated per nmole. A higher level of mutation is caused by the chemical agent aflatoxin, a toxin produced by a fungus. It induces 7057 mutants per nmole.

a. What quantity of epoxybutane would be required to equal the mutant-causing ability of 1 nmole of aflatoxin?

b. What assumptions did you make?

11-3 Mutations occur in chloroplasts (cp) and mitochondrial (mt) DNA as well as in nuclear (nuc) DNA. Nuclear DNA point mutation rates are similar in plants and animals. The plant mtDNA rate is less than 1/3 of the cpDNA rate, which is 1/2 the nucDNA rate. Animal mtDNA mutates 5 times faster than nucDNA.

a. How much faster or slower is the mutation rate of plant mtDNA compared to animal mtDNA?

b. What assumptions did you make?

11-4 Naturally occuring mutation rates have been determined for unicellular organisms, plants, and animals. Mutations can be caused by environmental factors, but also by mutator genes that can alter DNA replication and repair as well as by gene jumping—transposable elements.

Corn endosperm gene mutation rates due to transposable elements vary from a high frequency of $p = 0.00049$ to a low frequency of $p = 0.000001$.

If the frequency, or probability, of a mutant kernel is p, the probability of a nonmutant kernel is $q = 1 - p$. In a sample of n kernels, the probability of no mutants is q^n. The probability of at least one mutant is $1 - q^n$.

a. At each of the two mutation rates, how many kernels would we need to inspect to be 90% certain of seeing at least one mutant kernel?

b. What assumptions did you make?

11-5 One thing biologists do in order to determine how a cell or organism works is to find a mutant that does not do the activity. They go on a mutant hunt. Because this can take a long time, biologists speed up the process by using mutagens—often chemicals or radiation.

If the frequency, or probability, of a mutant cell is p, the probability of a nonmutant cell is $q = 1 - p$. In a sample of n cells, the probability of no mutants is q^n. The probability of at least one mutant is $1 - q^n$.

A specific adenine mutation in fungal spores occurs at different rates with different treatments. (Rates below have been multiplied by 10^6.)

Treatment	Rate
Natural conditions	0.4
EMS (1% ethylmethane sulfate) 90 min duration	25
UV rays (600 erg/mm^2/min) 6 min duration	375

a. Under each of these conditions, how many spores would you need to observe in order to be 90% certain of finding at least one mutant?

b. What assumptions did you make?

11-6 When mutations are induced by mutagens, some cells are killed and some of the survivors are mutated. If 1 million *Neurospora* spores are exposed to X-rays (2000r/min for 18 min), 84% of the spores are killed, and of the survivors, 259 evidence mutation.

If p is the probability of a mutant spore, the probability of a nonmutant spore is $q = 1 - p$. In a sample of n spores, the probability of no mutants is q^n. The probability of at least one mutant is $1 - q^n$.

a. What percentage of the surviving spores evidence a mutation?

b. How many of the surviving spores would need to be checked to be 90% certain of finding at least one mutant?

c. How many would you need to check to be 95% certain?

d. What assumptions did you make?

Protein Synthesis and Lifetime

Answers to Chapter 12 begin on page 232.

12-1 Protein synthesis requires both ATP and GTP (its energy equivalent). When the mononucleotides AMP or GMP are regenerated to ATP or GTP, two ATPs must be dephosphorylated to ADP molecules. For example,

$$GMP^* + ATP \longrightarrow GDP^* + ADP$$

$$GDP^* + ATP \longrightarrow GTP^* + ADP$$

Thus as proteins are made, respiration is required to continue to provide the needed ATP and GTP.

During *transcription*, each nucleotide added to a growing strand of mRNA requires an ATP equivalent. Both "high-energy" bonds are broken in the process.

During the *attachment* of amino acids to their tRNA, each amino acid added requires an ATP. Again both "high-energy" bonds are used.

During *translation*,

• Initiation requires 1 GTP

• Elongation requires 2 GTPs per codon step

• Termination requires 1 GTP.

To make one human hemoglobin molecule (146 amino acids long),

a. How many ATPs are required during the transcription phase?

b. How many ATPs are required during the attachment phase?

c. How many ATPs are required in total?

d. How many GTPs are required to make one human hemoglobin molecule?

e. What assumptions did you make?

12-2 In one mammalian cell it is estimated that 10,000 to 20,000 different types of mRNA can be found. Abundant mRNAs exist in many copies per cell (up to 12,000 copies/cell); scarce mRNAs (5 to 15 copies/cell) also can be detected. At any instant, a snapshot of mRNA content would reveal a total of 360,000 mRNAs. The cell usually has about 10 times as many ribosomes as mRNAs. Assume 75% of the ribosomes are involved in protein synthesis at any instant in time.

a. What is the maximum number of proteins that could be in the process of synthesis at any instant?

b. How many scarce proteins (use 5 copies/cell) could be in the process of being made?

c. How many abundant proteins could be in the process of being made?

d. What is the ratio of developing abundant proteins to developing scarce proteins?

e. What assumptions did you make?

12-3 Protein molecules in the cytosol of a cell have different half-lives $t_{1/2}$. The half-life is the time needed for 50% of the molecules to be lost or altered. Half-lives are determined by several factors, one of which is the "marking" of the protein by ubiquitin, which signals that the protein is intended for digestion by proteases. Ubiquitin binds differently to proteins due to differences in their amino ends. The following table gives the half-lives $t_{1/2}$ for some proteins with different amino ends:

Amino Acid End of Protein	$t_{1/2}$
MET (methionine)	>20 hrs
SER (serine)	>20 hrs
THR (threonine)	>20 hrs
ALA (alanine)	>20 hrs
VAL (valine)	>20 hrs
LEU (leucine)	<3 min
PHE (phenylalanine)	<3 min
ASP (asparagine)	<3 min
LYS (lysine)	<3 min
ARG (arginine)	<2 min

Half-life is related to the rate K of protein loss by the equation

$$t_{1/2} = \frac{0.693}{K}$$

where 0.693 is the natural logarithm of 2 and K is measured in reciprocal time units.

a. What is the rate of protein loss per minute when MET is the amino terminal? (Assume $t_{1/2} = 24$ hr.)

b. Compare with the rate when ARG is the amino terminal end. (Assume $t_{1/2} = 2$ min.)

c. What percentage of the protein with a MET amino end exists in the cell after 5 half-lives?

d. What percentage of the protein with an ARG amino end exists in the cell after 5 half-lives?

e. How much faster (as a percentage) is the rate of ARG-amino-end protein degradation as compared to MET-amino-end degradation?

f. Plot the loss of a population of protein molecules that have a MET amino end. Use time t on the horizontal axis and size of the population on the vertical axis, starting with 100% of the population at the starting time $t = 0$.

g. On the same graph, plot the loss of a protein with ARG as the amino terminus.

h. Hemoglobin exists in the cytoplasm of a red blood cell (RBC). Red blood cells last about 120 days in the bloodstream. In your initial experiments you find that the amino terminus of one protein chain in RBCs is either valine or leucine (a difference of only one methyl group). Which is more likely to be the correct amino terminus of this protein? (Assume $t_{1/2} = 24$ hr for valine and 2 min for leucine.)

i. What assumptions did you make?

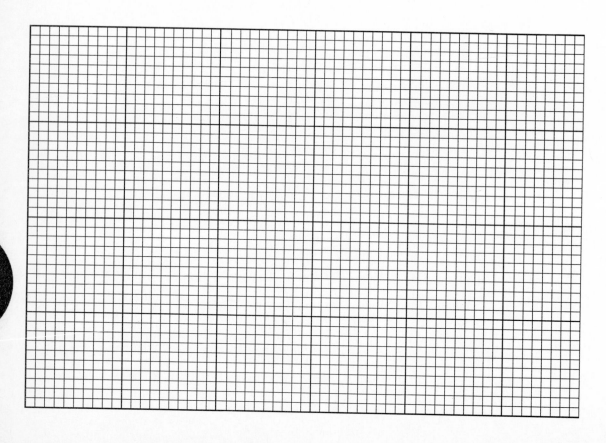

12-4 Assume you had 3 copies per cell of mature mRNA for a hemoglobin polypeptide chain of 145 amino acid units. One ribosome can attach onto an mRNA every 80 nucleotides. One amino acid can be translated in 60 milliseconds.

a. How many hemoglobin polypeptides could be made in 3 minutes?

b. What assumptions did you make?

CHAPTER 13

Biotechnology

Answers to Chapter 13 begin on page 235.

13-1 Human DNA has about 6 billion base pairs (bp). If the entire genome were to be cloned in vectors, the DNA would need to be cut into segments of the appropriate length, depending on the vector used. For some commonly used vectors, the insert size in kilo base pairs is

Vectors	Insert Size (kilo base pairs)
Bacterial plasmids	15
Phages	5–25
Cosmids	35–45
YACs	200–300

where YACs are yeast artificial chromosomes.

a. How many vectors of each type are needed to generate a human genomic library?

b. What assumptions did you make?

13-2 Here are two samples of double-stranded DNA fragments. The oligonucleotide sequence is known for each.

5′ AAGCCTTTAGCC3′   5′ GGGCTAGCTGCC3′

[complimentary strand]   [complimentary strand]

a. How many H bonds hold the strands in the first fragment together?

b. How many H bonds hold the strands in the second fragment together?

c. Heating will cause H bonds to break. Which fragment will become single-stranded more easily?

d. What assumptions did you make?

13-3 You have determined the amino acid sequences of two short segments of a protein of interest to you.

Segment A is the sequence   Gly–Val–Arg–Pro–Leu–Ser

Segment B is the sequence   Trp–Glu–Met–Asn–Gln–Trp

You decide to use this amino acid sequence information to generate a DNA sequence to be used as a probe (this can be done by a gene machine) to select a RFLP from your library. Using the codon matrix on the next page,

a. How many different DNA sequences could code for segment A?

b. How many different DNA sequences could code for segment B?

c. Which segment would you use?

d. What assumptions did you make?

		SECOND BASE			
	U	**C**	**A**	**G**	
U	UUU ⎤ Phe UUC ⎦ UUA ⎤ Leu UUG ⎦	UCU ⎤ UCC UCA Ser UCG ⎦	UAU ⎤ Tyr UAC ⎦ UAA Stop UAG Stop	UGU ⎤ Cys UGC ⎦ UGA Stop UGG Trp	U C A G
C	CUU ⎤ CUC CUA Leu CUG ⎦	CCU ⎤ CCC CCA Pro CCG ⎦	CAU ⎤ His CAC ⎦ CAA ⎤ Gln CAG ⎦	CGU ⎤ CGC CGA Arg CGG ⎦	U C A G
A	AUU ⎤ AUC Ile AUA ⎦ AUG Met or start	ACU ⎤ ACC ACA Thr ACG ⎦	AAU ⎤ Asn AAC ⎦ AAA ⎤ Lys AAG ⎦	AGU ⎤ Ser AGC ⎦ AGA ⎤ Arg AGG ⎦	U C A G
G	GUU ⎤ GUC GUA Val GUG ⎦	GCU ⎤ GCC GCA Ala GCG ⎦	GAU ⎤ Asp GAC ⎦ GAA ⎤ Glu GAG ⎦	GGU ⎤ GGC GGA Gly GGG ⎦	U C A G

FIRST BASE (5' end) — THIRD BASE (3' end)

PART IV EVOLUTION

CHAPTER 14

Evolution

Answers to Chapter 14 begin on page 237.

14-1 Natural selection over an extended period of time can cause one homogeneous population to devolve into two. When humans intervene, they can select desirable traits and breed different kinds of plants and animals much more quickly.

At the University of Illinois, a population of corn seeds averaged 10.9% protein by weight. Seeds with high protein and seeds with low protein were separated out. The high-protein seeds were crossed with high-protein seeds and the low protein were crossed with low protein for 50 generations. At the end, the high-protein progeny averaged 19.4% protein and the low averaged 4.9%.

a. Which trait, high or low protein, was modified more because of selection?

b. What percentage increase in the high-protein population and decrease in the low-protein population were accomplished after 50 years?

c. What were the average rates of increase and decrease in protein per year for the high- and low-protein populations?

d. What assumptions did you make?

14-2 This problem concerns the genes you have inherited from your ancestors.

a. What is the likely percentage of your DNA that came from your father's mother's father?

b. What is the theoretical maximum amount you could have inherited?

c. If the average length of a generation is 25 years, how many different people since 1776 have contributed to your DNA? How many since 1066?

d. What assumptions did you make?

14-3 Fossils found in rock suggest that

Life started on earth	3,500 million years ago
Plants live on land	400 million years ago
Flowering plants (monocots) "develop"	80 million years ago
Dinosaurs "vanish"	65 million years ago
Humans "appear"	150,000 years ago

Compare these time spans by changing the scale from years to seconds: let 1 yr = 1 sec. Thus, in this scale the average biology student is 20 seconds old.

a. In this scale, how many days ago did humans appear?

b. How many years ago did the dinosaurs vanish?

c. How many years ago did flowering plants begin to dominate?

d. How many years ago did plants start to live on land?

e. How many years ago did cells first fossilize?

f. What assumptions did you make?

14-4 In studying evolution, a common characteristic can be compared to determine how much variability exists among species. An example in vertebrate species is the number of differences in hemoglobin. Hemoglobin has 146 amino acids.

The following chart compares hemoglobin from five other species to that of humans:

Species	Number of Different Amino Acids	Estimated Millions of Years from Common Ancestor
Rhesus monkey	8	26
Mouse	25	80
Chicken	45	275
Frog	70	330
Lamprey	125	450

a. Find the average number of years per mutation in hemoglobin for each species listed.

b. Find the mutation rate in mutations per million years for each species.

c. Which animal shows a mutation rate closest to the average?

d. What percentage of the amino acids are different from humans in the monkey and the lamprey?

e. What assumptions did you make?

Chapter 14

14-5 Modern humans have an average brain volume of 1400 cubic centimeters; the range is from 1200 to 1800. Brain size can be traced in human evolution.

Humanoid History	Million Years Before the Present	Brain Capacity (in cm³)
Australopithecus	3–2.4	500–700
Homo habilis	2.4–1.6	500–800
Homo erectus	1.9–0.5	800–1100
Homo sapiens (Neanderthal)	0.12–0.06	1400
Homo sapiens (Cro-Magnon)	0.035	1400
Homo sapiens	0.10–present	1400

Let (x_i, y_i) represent the ith data point; for example, $x_1 = -2.7$ (the average of -3 and -2.4) and $y_1 = 600$ (the average of 500 and 700).

a. Plot the data.

There is a linear function, $y = mx + b$, whose graph is the straight line that best fits these data. This line is called a *regression line*. If there are n data points, the formulas for the slope m and y-intercept b of the regression line are

$$m = \frac{(\sum x_i)(\sum y_i) - n(\sum x_i y_i)}{(\sum x_i)^2 - n\sum x_i^2}$$

$$b = \frac{1}{n}\left(\sum y_i - m\sum x_i\right)$$

b. Compute and plot the regression line.

c. On average, how much time did it take to add 100 cubic centimeters to the human brain?

d. Assuming that each of the species above evolved from the one listed before it, and assuming punctuated equilibrium occurred in humanoid history, what was the percentage increase in brain size from group to group above?

e. What assumptions did you make?

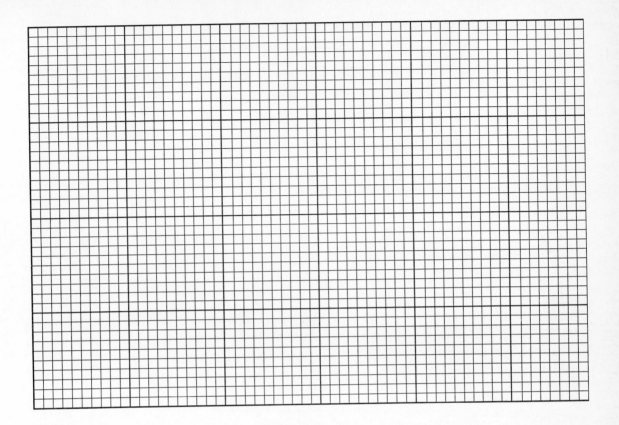

14-6 Estimating evolutionary time periods by employing gene product differences generates a continuum of very slow changes to relatively rapid changes. Nuclear gene products change (mutate) slower than mitochondrial gene products, which can change 5 to 10 times as fast. For this reason, mitochondria are often used when comparing closely related species.

One of the enzymes in mitochondria, NAD dehydrogenase, is mitochondrially encoded. When the first 75 amino acids in NAD dehydrogenase are compared between humans, gorillas and orangutans, the data in the table are obtained. Note that nucleotide differences can occur that do not necessarily cause a difference in the amino acid, since the code is degenerative.

Comparing	Amino Acid Differences	Nucleotide Differences
Human and gorilla	5	13
Human and orangutan	8	19
Gorilla and orangutan	9	26

Chapter 14

a. What percentage of the amino acids demonstrate differences between human and gorilla, human and orangutan, and gorilla and orangutan?

b. What percentage of the nucleotides demonstrate differences between human and gorilla, human and orangutan, and gorilla and orangutan?

c. Do gorillas or orangutans have NAD dehydrogenase more similar to that of humans?

d. Is the gorilla enzyme closer to that of humans or of orangutans?

e. The gorilla and human common ancestor separated from the orangutan ancestor about 10.5 million years ago. Gorillas and humans had a common ancestor about 10 million years ago. How many years elapse on average for each mutation in amino acids? Nucleotides?

f. What assumptions did you make?

14-7   Amino acid changes in proteins are used as an evolutionary clock. Different proteins mutate at different rates. One "unit evolutionary time" is the average time required for an acceptable amino acid change in a 100 amino acid sequence in a protein. Observed units of evolutionary time for four proteins are given below.

Protein	Unit Evolutionary Time (millions of years)
Fibrinopeptide	0.7
Hemoglobin	5
Cytochrome c	21
Histone 4	500

Histone 4 is 102 amino acids long in both peas and cows. There are two amino acid differences.

Cytochome c varies in size from 104 to about 112 amino acids. No differences exist between humans and chimpanzees and only one difference between humans and rhesus monkeys. There are 13 differences between humans and dogs, 20 between humans and rattlesnakes, 31 between humans and tuna, and 43 between humans and mung beans.

a. Using histone 4, how many years ago was there a common ancestor of peas and cows?

b. Using cytochome c, how many years ago was there a common ancestor of humans and each of the other animals and the plant listed?

c. How much faster does cytochrome c change compared to histone 4?

d. What assumptions did you make?

14-8 One evolutionary idea is that the number of amino acid differences in a similar protein from different species can be used as an evolutionary clock. The rate of amino acid substitution varies among proteins.

Protein	Changes per Protein per Billion Years
Fibronectin	9.0
Phospholipase A_2	1.9
Cytochrome b_5	0.45
Lactate dehydrogenase	0.34
Cytochrome c	0.22
Histone 4	0.01

a. What is the average rate of amino acid substitution in the proteins shown?

b. On average, how long would it take to make one change in histone H_4?

c. In the time it takes for one change in cytochrome b_5, how many changes would we expect to find in fibronectin?

d. In the 2 million years' time that horses have been differentiated from donkeys, how many amino acid substitutions would be found altogether in these six proteins?

e. What assumptions did you make?

14-9 The age of the earth is estimated from the amount of uranium isotope 238 remaining. Estimates are that only half of the U^{238} that was originally present still exists. Since the half life $t_{1/2}$ of U^{238} is 4.5×10^9, that puts the age of the earth at 4.5 billion years.

The earliest fossil evidence of bacteria is 3.5 billion years old. Vascular plants began to appear 400 million years ago. Modern humans are estimated to have appeared about 200,000 years ago.

a. What percentage of the original amount of U^{238} still remained at the time bacteria first became fossilized?

b. What percentage of the original amount of U^{238} still remained at the time vascular plants appeared?

c. What percentage of the original amount of U^{238} still remained at the time modern humans appeared?

d. What assumptions did you make?

14-10 Evolution occurs in jumps. An *E. coli* culture was maintained for 10,000 generations over 4 years. The liquid medium was changed daily to maintain a constant environment.

The average size of the cells at the start of the experiment was 0.35×10^{-15} liter. After 300 generations the size increased to 0.48×10^{-15} liter, and after another 300 generations the average increased to 0.49×10^{-15} liter. After 1200 generations it increased to 0.58×10^{-15} liter and remained so until the end of the experiment.

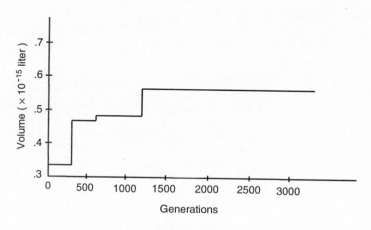

E. coli Evolution

a. How many hours long is a generation for *E. coli*?

b. What was the average size change over the course of the experiment?

c. What was the average size change in the first 300 generations?

d. How fast did the size increase over the course of the experiment in liters per generation?

e. How fast did the size increase over the course of the experiment in liters per year?

f. How fast did the size increase in the first 300 generations in liters per generation?

g. How fast did the size increase in the first 300 generations in liters per year?

h. What assumptions did you make?

14-11 New species arise by several different mechanisms. *Allopatric* speciation is due to changing geographical barriers, such as widening oceans in geological time. *Sympatric* speciation is due to a radical change in the genome. *Recombinant* speciation can occur naturally or because of human selection of desirable traits, and results in the many species of tulips and dogs that exist.

An example of allopatric speciation occurred with the sycamore. Sycamores in Eurasia and North America were separated about 50 million years ago. During that time mutations caused two identifiable species, although when Europeans brought the North American species to England in the 1600s the two species were crossed and produced a fully fertile hybrid.

Sympatric speciation occurred in goat's beard in less than 100 years (100 generations), by allopolyploidy.

In sunflowers, a "new species" can arise in 5 generations (5 years) by recombinant speciation.

   a. If the length of a generation of sycamores is 25 years (time from seed to seed), how many generations elapsed in the creation of the two morphological species?

   Of the examples listed above,

   b. How many times faster is recombinant speciation than the other two types, in terms of years?

   c. How many times faster is recombinant speciation than the other two types, in terms of generations?

   d. What assumptions did you make?

14-12 Geologists now believe that the surface of the earth is made up of large plates that slowly move around. Lighter material rides on top; this lighter material makes up the continents. At one point in time, all the land material had been collected together in one large continent called *Pangea*. Then about 200 million years ago, motion of the plates caused it to break up into the continents we know today.

Currently the North American continent is moving away from the European continent at the rate of 2.3 centimeters per year. The distance from New York to London is 5583 kilometers.

   a. If this rate of motion were constant since the breakup, how far would Europe be from North America?

   b. What would be the average speed necessary to have separated North America from Europe the current distance?

   c. As the continents separated, plants and animals became isolated. At the time of the breakup, cycads were dominant. Flowering plants appeared 135 million years ago. How far was North America from Europe then?

   d. What assumptions did you make?

Chapter 14

14-13 Extinction of species is very common over evolutionary time. Several tree species have survived over unexpectedly long time periods. These "living fossils" include the gingko, whose fossil record goes back 150 million years, the dawn redwood (fossil record 90 million years), and the wollenia pine (fossil record 200 million years).

The contemporary rate of average continental erosion is 0.03 millimeters per year. The cutting of the Grand Canyon is taking place at 0.7 millimeters per year.

a. If the contemporary rate of Grand Canyon cutting existed throughout the history of the dawn redwood, how many miles deep would the canyon be today?

b. The durations of the species above can be related to the human species, which developed 200,000 years ago. Assume that each of the trees listed has a 20-year-long generation (seed to seed). How many generations of each tree existed before and how many since 200,000 years ago?

c. What assumptions did you make?

PART V PLANT SCIENCE

CHAPTER 15

Structure

Answers to Chapter 15 begin on page 243.

15-1 During development, at one stage a plant embryo is an eight-celled sphere on a stalk. This sphere is measured to be 1 cm in diameter from a photograph known to be enlarged 435 times. At a later stage, the clump of cells is still spherical but now the diameter is 2.5 cm (also enlarged 435×). The individual cell size has not changed—only the number of cells.

Plant Embryo

a. How many cells are present in the 2.5 cm sphere?

b. What assumptions did you make?

15-2 The root meristem has a quiescent center where cells divide at a much lower rate than those that produce root cap cells. Root cap cell precursors divide every 13 hours, while the quiescent center cells divide every 430 hours.

a. How many times will the root cap cell precursor divide for each quiescent cell division?

b. How many cells can one root cap precursor produce for each cell division in the quiescent center?

c. What assumptions did you make?

15-3 The cell cycle in apical meristems has a duration as short as 20 hours. The cell cycle in the cambium (a lateral meristem) may be much longer. Cytoplasmic division is controlled by means of a phragmoplast (microtubules that orient Golgi-generated vesicles). The phragmoplast is shaped like a ring that migrates (grows) from the site of the chromosomal alignment at metaphase outward to the cell membrane. The growth of the diameter of the phragmoplast is between 50 and 100 μm/hr.

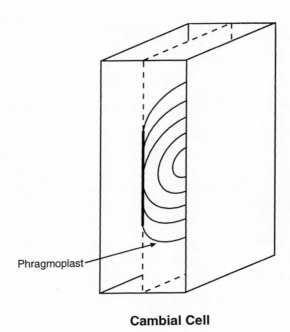

Phragmoplast

Cambial Cell

If the cambial cell has the dimensions of 25 μm by 25 μm by 8700 μm and we take the growth of the diameter of the phragmoplast to be 75 μm/hr,

    a. How long does it take for the growing phragmoplast to reach the "edge" and begin to split the cell?

    b. How long does it take for the cambial cell to become two new cells?

    c. How many times longer does it take to divide a lateral meristem as compared to an apical meristem?

    d. What assumptions did you make?

15-4 Shoot apical meristems have a central zone where cells divide more slowly than in the periphery (peripheral zone). In actively dividing meristems the range of time, in hours, for one cell division varies across species.

Species	Central Zone	Peripheral Zone
Rapid dividers	>40	30
Slow dividers	288	157

a. How much more rapid is the rapid divider when compared to the slow divider?

b. How much more rapid is the peripheral zone division when compared to the central zone?

c. In each case, how many days does it take an apical cell to divide 10 times?

d. What assumptions did you make?

15-5 Water is moved in tall plants through pipes made up of cells. In some plants these cells are *tracheids* and in others *vessel elements*. Tracheids are narrow and have end walls with holes called pits. Vessel elements are wider and their end walls have larger holes (perforations), or end walls may be absent. Thus, water moves more readily through vessel elements. *Vessels* are long pipes of stacked vessel elements. They vary in size due to genetics and environmental conditions during their growing stages, but can be as long as 50 feet in some oak trees. Individual vessel elements vary in length from 50 to 100 μm; tracheids can be up to 1 mm in length.

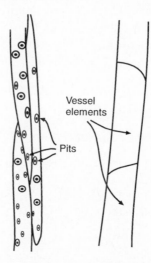

Vessel elements

Pits

Tracheids   **Vessel**

a. How many cells make up a 50-foot oak vessel?

b. If water were to pass through a tracheid pathway the same 50 feet, how many end walls would need to be traversed?

c. What assumptions did you make?

15-6 Depending upon the species, the number of stomates on the surface of a leaf varies from 1000 to 60,000 stomates per square centimeter. The size of the stomatal opening varies also. Generally, the opening appears oval with the aperature from 3 to 12 μm in the short dimension and 10 to 40 μm in the long dimension.

We can approximate the oval by an ellipse. The area of an ellipse is given by $A = \pi ab$, where a is the *minor semiaxis* (one-half the shortest diameter) and b is the *major semiaxis* (one-half the longest diameter).

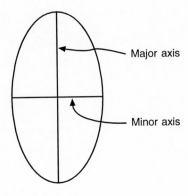

a. What percentage of the leaf area is made of stomatal openings?

b. What assumptions did you make?

15-7 Corn (*Zea mays*) has about 6000 stomates per square centimeter on the upper epidermis and about 10,000 stomates per square centimeter on the lower epidermis. Transpirational loss of water from corn varies due to varietal differences and environmental conditions, but a representative value would be 700 grams per square meter per hour.

a. What is the number of water molecules lost per stomate in one minute?

b. What assumptions did you make?

15-8   A young sugar maple tree in your backyard has a circumference of 15 cm. Because of an injury during mowing, you know that the bark is 0.5 cm thick. After 10 years, you tap the tree to collect sap for maple syrup. The plug you remove has ten annual rings of wood that average 0.6 cm in thickness.

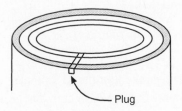

Plug

Sugar Maple Tree

a. Using the formula for the area of a circle, find a formula for the area of a cross-sectional ring in terms of its outer radius and its inner radius.

b. Calculate the cross-sectional area of the wood that was produced during each of the 10 years.

c. Calculate the percentage increase in the area of the ring each year.

d. Plot the yearly increase. Is the graph straight or curved?

e. What assumptions did you make?

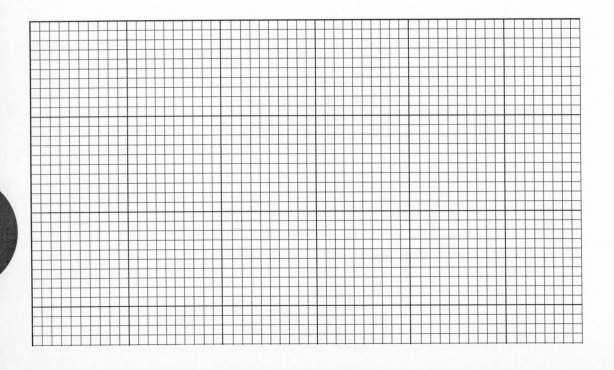

15-9 Palm trees have no cambia, therefore all above-ground cells are derived from an apical meristem in the stem tip. The circumference of a palm tree is 2.5 feet. The cell dimensions are on average 50 μm by 50 μm by 100 μm (the long axis being vertical).

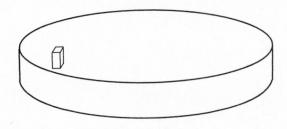

Palm Tree Slab

a. How many cells are exposed in a cross section of the palm tree?

b. If a 4-inch-thick slab is used to make a stool, how many many cells does it contain?

c. What assumptions did you make?

15-10 *Plasmodesmata* are cytoplasmic connections across plant cell walls that connect adjacent cell cytoplasms. Some cells have few plasmodesmata connections while others have more; this is due to genetics, age, and location within the plant.

The density of plasmodesmata within the cell membrane ranges from a high of 25 per square micrometer to a low of 0.2 per square micrometer. The average plasmodesmata tube is 40 nm in diameter.

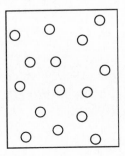

a. What percentage of the cell membrane surface area is composed of plasmodesmata at the high density?

b. What percentage of the cell membrane surface area is composed of plasmodesmata at the low density?

c. What assumptions did you make?

15-11 If a potted plant is knocked over, after some time the shoot tip will bend and begin to grow again in a vertical orientation. The bending is due to unequal elongation of the upper and lower cells in the plant organ. The diagram shows the curvature seen in a sunflower shoot.

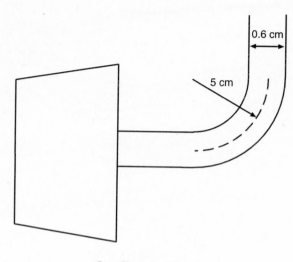

Sunflower Shoot

a. What are the radii of the inner and outer arcs made by the upper and lower sides of the stem?

b. What are the lengths of the inner and outer arcs?

c. What is the ratio of the length of the inner arc to that of the outer arc?

d. If the cells average 100 μm in length, how many cells make up the lengths of the inner arc and the outer arc?

e. In fact, although the cells *average* 100 μm, microscopic observation reveals that there is actually the *same number* of cells on the inner arc as on the outer arc. What is the length of a cell on the inner arc? the outer arc?

f. What assumptions did you make?

Long-Distance Transport

Answers to Chapter 16 begin on page 248.

16-1 Liquid water moves into and out of cells by diffusion. Water vapor moves from the inside of the plant leaf to the ambient air by diffusion, a process known as *transpiration*. This transpiration causes dissolved minerals to be moved long distances in the plant. The time t in seconds for water to move a distance d in meters is given by

$$t = \frac{d^2}{D}$$

where D is the *diffusion coefficient*. A reasonable value for D is 2.4×10^{-5} m^2/sec. (NOTE: Water vapor diffuses in air much more rapidly than in liquid water.)

The path of the diffusion of water vapor from a leaf into the air varies considerably, but a measured distance of 1 mm is reasonable.

  a. How long does it take a molecule of water vapor to be lost by this leaf?

    Hairiness of leaves is a genetic trait. Leaf hairs may double the distance water must diffuse.

  b. How long would it take to lose water from a hairy leaf?

  c. What assumptions did you make?

16-2 Diffusion is rapid over very short distances and slow over longer distances. The time t in seconds required for diffusion is given by

$$t = \frac{d^2}{D}$$

where d is the distance in meters and D is the *diffusion coefficient*. A reasonable value for D for a small molecule is 10^{-9} m^2/sec.

  a. How long would it take for a molecule to diffuse across a 50 μm cell?

  b. How long would it take for a molecule to diffuse across the 1-m length of a corn leaf?

  c. What assumptions did you make?

16-3 Nitrogen fixation in the biological world is accomplished primarily by microorganisms, acting either alone or in combination with plants. Nitrogen in the air can be made into ammonia by an enzyme complex called *nitrogenase*. The overall reaction of nitrogenase is

$$N_2 + 8 \text{ NADPH} + 16 \text{ ATP} \longrightarrow 2 \text{ NH}_3 + 8 \text{ NADP}^+ + 16 \text{ ADP} + 16 \text{ P}$$

Nitrogen fixation can be compared to carbon fixation in photosynthesis. The overall reaction of photosynthesis is

$$CO_2 + 2 \text{ NADPH} + 3 \text{ ATP} \longrightarrow C_{\text{fixed}} + 2 \text{ NADP}^+ + 3 \text{ ADP} + 3 \text{ P}$$

The energy released when

NADPH $\longrightarrow$ NADP$^+$ is $\Delta G = -51.7$ kcal/mol;

ATP $\longrightarrow$ ADP is $\Delta G = -7.3$ kcal/mol.

a. What is the energy cost of making ammonia?

b. What is the energy cost of carbon fixation?

c. How much more energy-expensive is nitrogen fixation than carbon fixation?

To convert NH_3 to an amino acid, one possible pair of reactions is the following:

$NH_3 +$ glutamic acid $+$ ATP $\longrightarrow$ glutamine $+$ ADP $+$ P

glutamine $+$ NADPH $+ \alpha$ketoglutaric acid $\longrightarrow$ 2 glutamic acid $+$ NADP$^+$

These two reactions can be summarized in

$NH_3 +$ ATP $+ \alpha$ketoglutaric acid $+$ NADPH $\longrightarrow$ glutamic acid $+$ ADP $+$ P $+$ NADP$^+$

d. What is the total energy required for the conversion of nitrogen in air to nitrogen in an amino acid?

e. What assumptions did you make?

16-4 Plot the water potential Ψ (measured in megapascals (MPa)) of air at 20°C vs. the relative humidity (RH). Because of the great variation in Ψ, semi-logarithm graph paper (on the next page) is best for this purpose.

RH (%)	Ψ (MPa)
100	0
99.5	−1.56
99.0	−3.12
98.0	−6.28
95.0	−15.95
90.0	−32.8
75.0	−89.4
50.0	−215.5
20.0	−500
10.0	−718

a. What is your estimate of Ψ at relative humidities of

30% _____
40% _____
60% _____
80% _____

b. What assumptions did you make?

16-5 In the soil–plant–air continuum, water moves from a higher water potential to a lower water potential. Soil water potential varies from

- field capacity (water potential about −0.01 MPa), where the soil is as wet as it can be after gravity removes excess water, to

- permanent wilting point (water potential about −3 MPa), where the soil is so dry that the plant is always wilted. This point varies with soil type and plant type.

Soil water is "held" by soil particles, forming a meniscus between soil particles. The equation that relates water potential to water menisci is

$$P = -2\frac{T}{r}$$

Chapter 16

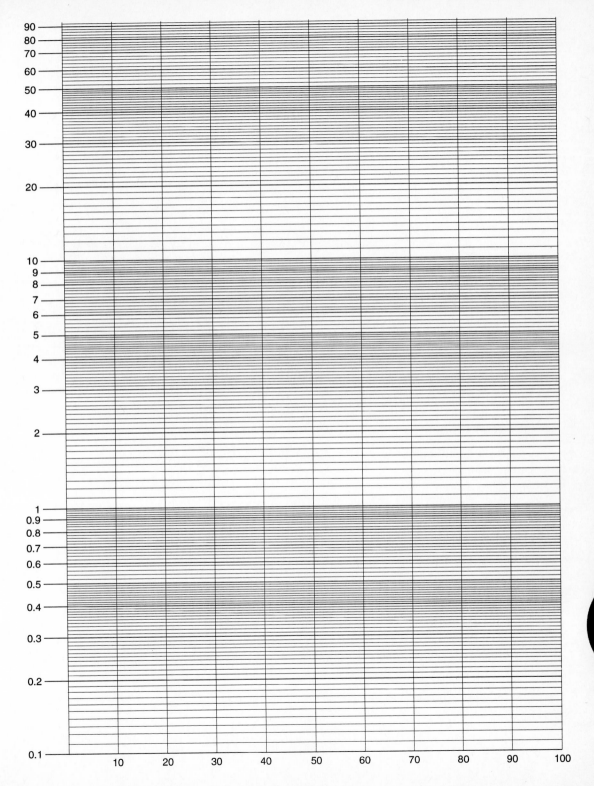

where

P = water potential (MPa)

T = surface tension of water (7.28×10^{-8} MPa · m)

r = radius of curvature of meniscus in meters

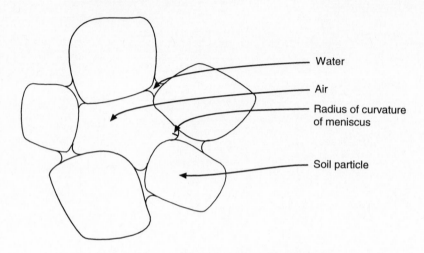

Water

Air

Radius of curvature
of meniscus

Soil particle

Meniscus Between Soil Particles

a. What is the functional average radius of curvature of the meniscus of soil water at field capacity?

b. What is the functional average radius of curvature at the permanent wilting point?

c. What assumptions did you make?

16-6 The internal surface area inside a leaf is 7 to 30 times the external surface area. Water evaporates in the internal space and diffuses out holes in the leaf called *stomata*. The internal evaporative surface consists of the cell walls and the menisci between adjacent cells. Water in the menisci is freer to evaporate. The water potential in leaves varies from around −0.3 MPa to around −3.0 MPa when leaves are in sunlight.

Water potential is related to menisci of water by the equation

$$P = -2\frac{T}{r}$$

where

P is the pressure in MPa

T is the surface tension of water (7.28×10^{-8} MPa·m)

r is the radius of curvature of the meniscus in meters

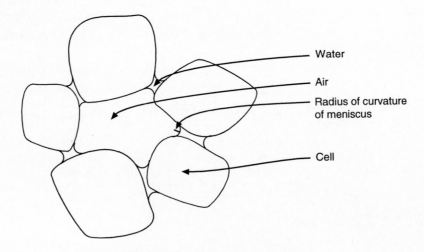

Meniscus Between Leaf Cells

a. What is the functional average meniscus radius in the leaf?

b. What assumptions did you make?

16-7 You mow grass for a summer job. Many of your customers wish for you to collect the clippings and take them with you. A typical lawn of about 1/3 acre produces 30 pounds of clippings for each mowing. About 80% of a cell's water is in the vacuole. The average vacuolar concentrations of three important mineral nutrients in grasses are:

	Ion	mM	Amount per Clipping	Punds per Acre
Nitrate	NO_3^-	45	_____	_____
Phosphate	HPO_4^-	65	_____	_____
Potassium	K^+	145	_____	_____

a. Fill in the fourth column with the amounts in moles of nitrate, phosphate, and potassium that you remove with each mowing (vacuolar contents only).

b. Fill in the fifth column with the pounds per acre of each.

c. What assumptions did you make?

16-8 In the development of a leaf, leaf cell division is at its highest rate in the early stages, and leaf elongation or growth takes over in the later stages. Meanwhile, the growth in the number of plastid genomes per plastid and in the growth of the number of plastids per cell also depends on leaf size.

The following table lists the number of copies of the plastid genome in each plastid of a spinach leaf at different times in the growth of the leaf. It also lists the number of plastids per cell and the percentage of DNA in the leaf that occurs in the plastids.

Leaf Length in (mm)	Copies of Plastid Genome per Plastid	Plastids per Cell	Plastid DNA as a Percentage of Total DNA	Plastid Genomes per Cell
1	76	10	7	_____
2	150	10	8	_____
20	190	29	23	_____
100	32	171	23	_____

a. Compute the last column of data, the number of plastid genomes per cell.

b. Plot each of the four sets of data using spinach leaf length along the x-axis.

c. When, in the leaf growth process, is the leaf making plastid DNA per cell most rapidly?

d. When, in the leaf growth process, is the leaf making plastids per cell most rapidly?

e. What assumptions did you make?

16-9 The increase in dry weight of a pumpkin was measured over the period of a month. In 792 hours the dry weight increased 482 grams. The area of a cross section of the phloem portion of the peduncle (stalk) of the pumpkin was 18.6 mm^2. Of this phloem area, only about 20% was sieve tubes, the pathway for dry weight movement to the fruit.

a. What was the mass transfer rate (grams of dry weight moved per mm^2 per hour)?

The density or specific gravity of the phloem contents was 1.5 g/cm^3. The velocity of dry weight movement can be determined by dividing the rate from (a) by the density.

b. What was the velocity of dry weight movement?

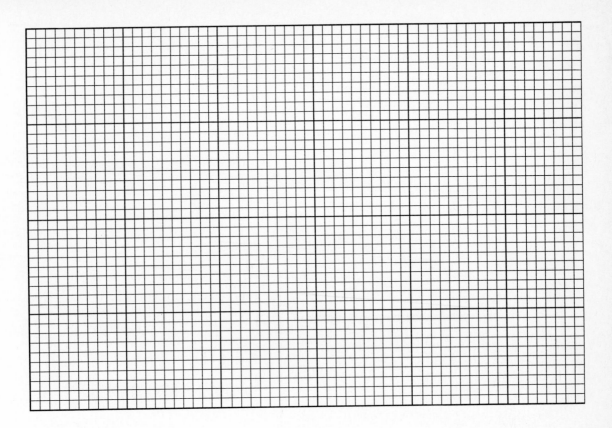

(Problem 16-9 continued)

The phloem contents in pumpkin are about 20% sucrose, so the velocity of sugar water movement is 5 times the value of (b).

c. What was the velocity of sugar water movement?

Published sugar water velocities for different plants vary from 40 to 290 cm/hr.

d. Is pumpkin a relatively slow or relatively fast sugar water mover?

e. What assumptions did you make?

CHAPTER 17

Plant
Development

Answers to Chapter 17 begin on page 253.

17-1 Light alters the length of the palisade (column-like) cells in a tree leaf. In the same plant, leaves growing in full sunlight are different from those growing in the shade of leaves above them. In a sun-grown leaf, the palisade cells compose about 1/2 the thickness. In a shade-grown leaf, the palisade cells are 2/3 as long as those in a sun-grown leaf, while the other cells have approximately the same thickness.

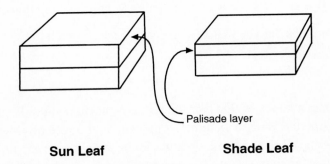

Palisade layer

Sun Leaf   **Shade Leaf**

a. How much thicker (as a percentage) are sun-grown leaves?

b. What assumptions did you make?

17-2 Around 80% of cells in a plant that become specialized make copies of DNA in their nuclei. Two mechanisms are involved: endoreduplication and gene amplification. Endoreduplication yields copies of the entire chromosome. Gene amplification copies parts of the chromosome. Endoreduplication yields DNA content that will be a power of 2 times the amount of DNA in a G_1 diploid cell. A list of some specialized cells and the number of copies of DNA in their nuclei are given below.

Specialized Cell Type	Copies of DNA	No. of S Cycles
Anther cells in a sedge	8	_____
Anther hairs in a cucumber	64	_____
Stinging leaf hairs in nettle	256	_____
Embryo cells (suspensor) in navy bean	2048	_____
Embryo cells (suspensor) in scarlet runner beans	8192	_____
Endosperm in *Arum*	24,576	_____

a. Fill in the third column with the number of times the differentiated cells have undergone an S cycle without an M cycle.

b. Are *Arum* endosperm nuclei genes amplified or endoreduplicated? Why?

c. What assumptions did you make?

Chapter 17

17-3 In *Arabidopsis*, a flower stalk is formed that can have many individual flowers. The meristem can make 1.9 flowers per day for 30 days. The flowers "open" (develop to *anthesis*) in 13.5 days.

  a. From the time of flower initiation until the first flower is open, how many flowers have been made?

  b. How many flowers are open when the meristem stops making flowers?

  c. What assumptions did you make?

17-4 One environmental factor that plants monitor is the duration of light. The length of an uninterrupted dark period is often extremely important in the kind of growth (vegetative as compared to flowering). Long-day (LD) plants require at least a certain day length before they begin to flower, whereas short-day (SD) plants require at most a certain day length before they begin to flower.

  In the corn belt in the United States, sunrise and sunset define the length of the day. In northern Iowa, the following times of sunrise and sunset on the first day of each month are presented below.

Month	Sunrise	Sunset
Jan	7:22	4:45
Feb	7:09	5:19
Mar	6:34	5:52
Apr	5:44	6:24
May	5:00	6:54
Jun	4:33	7:23
Jul	4:35	7:33
Aug	4:58	7:14
Sep	5:28	6:32
Oct	5:56	5:43
Nov	6:29	4:58
Dec	7:02	4:35

  a. Plot the sunrise and sunset times on graph paper.

  b. Plot the duration of light per day and the duration of dark per day on graph paper.

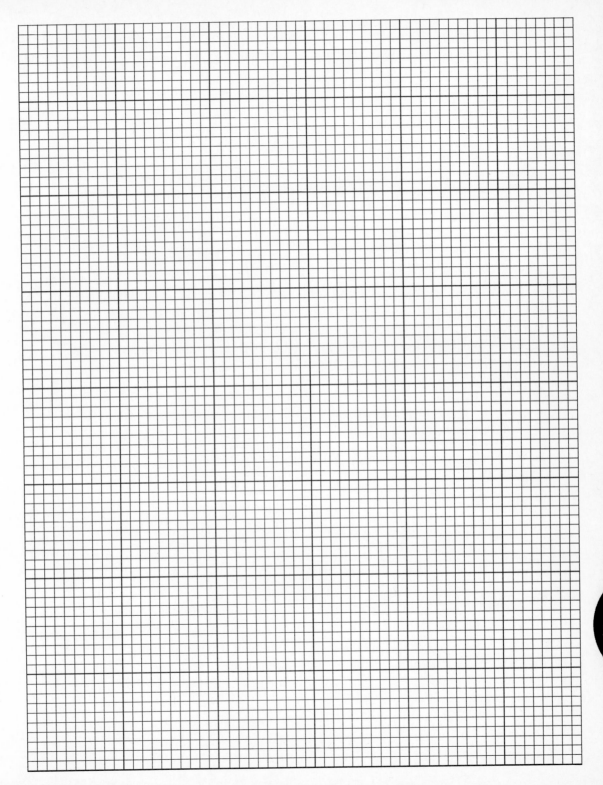

Suppose you have inherited a greenhouse. You don't want to pay for additional lighting, so you will use only sunlight. You have a local market for the following plants:

Plant	SD/LD	Critical Day Length
Dill	LD	11
Spinach	LD	13
Soybean	SD	15.5
Cocklebur	SD	14

c. When during the year would you expect each plant to be induced to flower?

d. What assumptions did you make?

17-5 Developing seeds of *Arabidopsis* produce cotyledon cells that are sites for storage of nutrients that are used later during germination.

At various stages in development (measured in hours-after-fertilization—HAF), cotyledon cell numbers and cell volumes increase. From 144 HAF to 216 HAF the cell number is constant (5400 cotyledon cells per seed), but the cell volume increases from 816 μm^3 to 1132 μm^3. Assume the cells are cubes.

a. What is the percentage increase in the volume of a cotyledon cell over the 72-hour period from 144 HAF to 216 HAF?

b. What is the percentage increase in the side length dimension of a cubical cell during this period?

c. What is the total storage volume per seed at the beginning and at the end of this period?

d. What assumptions did you make?

17-6 As a tree increases in diameter from year to year the vascular cambium must adjust. The vascular cambium is one cell thick and consists of two types of cells, *fusiform initials* and *ray initials*. The ratio of fusiform initials to ray initials is constant within a species. An average ratio would be about 25% of the cambial surface area composed of ray initials. When required, new ray initials are made from fusiform initials and additional fusiform initials (as the circumference increases) are made.

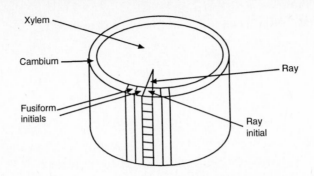

Cambial cell sizes vary from species to species, but an average fusiform cell is 8 μm by 8 μm by 200 μm (length varies from 140 μm to 8700 μm) and a ray initial is 8 μm by 8 μm by 8 μm.

You measure the xylem of a tree one year to be 20 cm in diameter and the next year to be 21 cm in diameter.

a. How many fusiform initials and ray initials surrounded the xylem the first year measured?

b. How many fusiform initials and ray initials surrounded the xylem the second year measured?

c. How many fusiform initials were converted to ray initials during that year?

d. What assumptions did you make?

17-7 You have decided to plant 1000 trees on 10 acres of land that you cannot farm. You plant a crop of walnut trees and tell your grandchildren they can harvest them when, on average, an 8-foot-long tree trunk has produced 6 board feet during the previous year. (A board foot of sawed wood is 12 in by 12 in by 1 in.)

Currently, the tree circumference is 6 inches and the average depth of the bark on the tree is 1/2 inch. The average increase in radius is 1/4 inch per year.

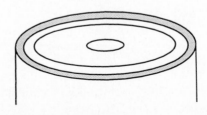

a. How many years will it be until the trees can be harvested?

b. What assumptions did you make?

17-8 *Arabidopsis* is a well-studied plant in the cabbage family. The first leaf of this plant is fully expanded 20 days after seed imbibition. When mature, the leaf area is around 30 mm^2 and the leaf thickness is about 100 μm. At this stage, the number of cells is 132,000. The cells can be grouped into three types:

Tissue Type	Percentage of Mature Leaf Volume	Percentage of Mature Cells
Mesophyll	60.5	36
Epidermis	12.5	36
Vascular	1.5	28

a. What is the average volume of a cell of each of the three types?

b. What is the volume of air internal to the leaf?

c. What assumptions did you make?

17-9 When *Arabidopsis* seedlings are grown in a nonsoil medium, root hair elongation can be studied. In normal seedlings, root hairs grow 100 micrometers per hour and grow to a length of 1.5 mm. The average diameter of a root hair is 0.7 mm. Golgi-derived vesicles (0.1 μm diameter) fuse with the cell membrane to enlarge the root hair.

a. How many vesicles are required to generate a root hair?

b. What is the final surface area of a root hair?

c. How long does it take to grow a root hair?

d. What is the rate of vesicle fusion to the root membrane during root hair formation?

e. What assumptions did you make?

17-10 In developing seeds of *Arabidopsis*, the endoplasmic reticulum (ER) is believed to play a role in the generation of stored nutrients. Both the volume and the surface area of ER show changes, as given in the table below. The greatest difference can be seen between hour 168 and hour 216 after fertilization. Assume the cell has the shape of a cube. The unit of length used in the table is micrometers.

Days	ER Volume per Cell	ER Surface Area per Cell	Cell Volume	Cell Surface Area
168	38	1413	951	_____
216	13	514	1132	_____

a. Compute the remaining column in the table.

b. What is the ratio of ER volume to cell volume at hours 168 and 216?

c. What is the ratio of ER surface area to cell surface area at hours 168 and 216?

d. What assumptions did you make?

17-11 The concentration of plant hormones varies from 10^{-6} to 10^{-9} M.

In a bioassay, the curvature of a coleoptile is correlated with the amount of the auxin IAA that can be released from an agar block. Maximum curvature occurs when the IAA concentration is 0.2 mg/l. The molecular weight of IAA is 180.

a. What is the concentration of IAA in moles per liter (M) in this agar block?

b. If the concentration of IAA in the elongating stem tip is 10^{-6} M and the apical meristem cells are cubes with 7 μm sides, what is the number of IAA molecules in the cell?

Elongating cells demonstrate a polar basipetal movement from cell to cell. The rate of movement is about 1 cm/hr.

c. Assuming that the elongating cells average 15 μm in length, how many cells does an IAA molecule traverse in one hour?

d. What assumptions did you make?

17-12 Fertilization of a flowering plant's egg requires pollen tube growth and the release of the sperm nuclei near the egg. Pollination and subsequent pollen tube growth through the style requires new cell membrane and cell wall production. The cell membrane and cell wall components are derived from Golgi and are delivered to the growing tube tip via 0.1-μm diameter vesicles moving along microfilaments.

Chapter 17

The rate of growth of the pollen tube varies depending on the type of plant and the environmental conditions (especially temperature). In some oaks, the time from pollination to fertilization is as long as 14 months. The quickest time reported from pollination to fertilization is 15 minutes.

Corn pollen tube growth has been measured to be 6.25 mm/hr. Corn silk is the corn style and can be 20 to 30 cm long. The corn pollen tube diameter is about 10 μm. The corn pollen tube cell wall is 20-nm thick.

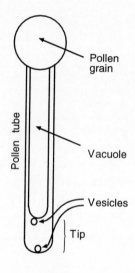

Corn Pollen Tube

a. How long does it take corn from pollination to fertilization?

b. How many vesicles are required to merge with the cell membrane to deliver the sperm nuclei?

c. What percentage of the volume of the vesicles is taken up by the cell wall components?

d. The growing pollen tube retains its cytoplasm in the tip 3 to 5 μm; what is the volume of the vacuole that is made in the corn pollen tube between the pollen grain and the tip at the time of fertilization?

e. What assumptions did you make?

CHAPTER 18

Comparative Physiology

Answers to Chapter 18 begin on page 260.

18-1 Goldfish obtain their oxygen from O_2 dissolved in water. They process water through their gills and may extract up to 80% of the dissolved O_2. When fish are active they of course use more O_2 than when they are inactive.

At low temperature (5°C), water contains 9 ml O_2 per liter and an active fish at this temperature uses 30 ml/kg/hr. At a high temperature (35°C), water contains 5 milliliters of O_2 per liter and an active fish at this temperature uses 285 ml/kg/hr.

   a. How much water in liters per kilograms per hour must be processed by an active fish at 5°C?

   b. How much must be processed by an active fish at 35°C?

   c. How many times more water must be processed by an active fish at 35°C than at 5°C?

   d. What assumptions did you make?

18-2 Animals vary in their metabolism as measured by oxygen consumption. Here are the data for some animals under standard conditions (low activity, darkness, quiet, habituation to experimental conditions):

Animal	Body Weight (grams)	ml O_2/g/hr
Rat	280	0.88
Cat	3,000	0.45
Dog	20,000	0.36
Man	71,500	0.21
Cow	500,000	0.124

   a. How long would it take each animal to consume a mole of O_2?

   b. How many milliliters of O_2 per minute will be consumed by each animal?

   c. What assumptions did you make?

18-3 Animals that grow larger while retaining their shape exhibit *isometric growth*. Arthropods (crabs, lobsters, grasshoppers, lice, etc.) are examples. Arthropods generally molt when they have doubled their weight. The volume of the animal is proportional to its weight. In addition, the volume is related to the length by Brook's law of linear growth: that volume is proportional to the cube of the length. Thus we obtain the formula that relates weight w and length l:

$$w = kl^3$$

where k is a constant of proportionality.

a. If an animal that grows isometrically doubles its length in a growth period, by how many times does the weight increase?

b. How many times longer does an animal need to grow in order to double its weight and molt?

c. If an animal grows from 4 inches to 5 inches, how much would its weight increase?

   An average human baby is about 20 inches long and weighs 7 pounds at birth. An average adult is 70 inches tall and weighs 150 pounds.

d. If humans grew isometrically, how much should a 70-inch-tall human weigh?

e. What assumptions did you make?

18-4 Flying for animals is energy-expensive. Some animals are more efficient than others in using oxygen for flight. Below are listed the respiration rates (in milliliters O_2 per gram per hour) of four animals at rest and when flying.

Animal	Rest	Flying
Butterfly	0.7	100
Fruit fly	1.68	21
Hummingbird	14	85
Pigeon	0.89	11.9

Human beings, clearly unable to fly, at maximal exercise use 20 times the oxygen they use at rest.

a. Which of the fliers has the least increase in O_2 consumption during flight compared to rest?

b. Which flier comes closest to humans in the increase in O_2 consumption during exercise?

c. What assumptions did you make?

Chapter 18

18-5   Paramecia move in water 1 millimeter per second, about 10,000 body lengths per hour.

   a. How long is a paramecium?

   b. How fast in miles per hour would a 6-foot-tall biology student have to swim to achieve a similar rate of movement?

   c. What assumptions did you make?

18-6   In hermaphroditic snails, male copulatory activity is eliminated by removal of a part of the vas deferens. After this operation the consumption of food, energy storage, and growth are the same for these now noncopulant snails; however, the egg-laying rate is increased. The still copulant snails that were the controls in this experiment received a sham operation.

   During one 28-day experiment, copulent snail average egg production was 25.6 eggs per day and noncopulant egg production rose to 40.8 eggs per day.

   a. How many eggs were produced during the experiment by both the copulant and noncopulant snails?

   b. How many eggs were used to provide the energy for copulatory behavior and sperm production in snails during this 28-day period?

   c. How many eggs were used per day to provide the energy for copulatory behavior and sperm production?

   d. By what percentage was egg production increased by eliminating copulatory behavior?

   e. What assumptions did you make?

18-7   Cheetahs eat gazelles. Cheetahs stalk and strive to outrun gazelles quickly. Gazelles can run at 50 miles per hour. Cheetahs have a top speed of 71 miles per hour, but they can maintain that speed for only 20 seconds. After that a gazelle can outrun a cheetah.

   a. Assuming similar rates of acceleration, how much head start distance must a gazelle have to just escape the cheetah's charge?

   b. What assumptions did you make?

18-8 Insects rely on diffusion to exchange respiratory gases with the environment. They lack a system of lungs and a circulatory system. They do, however, contain microscopic air tubes that serve as areas of diffusional exchange. Insects are rarely over 3/4 inch thick. The time required for diffusion of CO_2 and O_2 can be calculated from

$$t = \frac{d^2}{D}$$

where t is time measured in seconds, d is distance in meters, and D is a constant equal to 2.4×10^{-5} square meters per second.

a. How long does it take to move O_2 to the center of the insect?

b. What assumptions did you make?

18-9 Fireflies signal potential mates by flashing—emitting light. In one kind of firefly the signal flashes are 1.1 to 1.7 seconds apart. Each flash of the firefly can emit 10^{14} to 10^{15} photons or quanta of yellow light, light at wavelength 560 to 580 nanometers. One molecule of ATP activates one molecule of luciferin to emit one photon of light.

a. How many flashes of light can 1 mole of ATP make?

b. Thirty-six molecules of ATP can be made from each glucose molecule. How many glucose molecules are needed per firefly flash?

c. What assumptions did you make?

18-10 A hen's egg is made from a released ovum that travels through the oviduct, gaining components along the way. These contibutions are

Oviduct Component	Average Length (cm)	Component	Amount (g)	Percent Solid	Time Spent (hours)
Infundibulum	11.0	chalazal			$\frac{1}{4}$
Magnum	33.6	albumen	32.9	12.2	3
Isthmus	10.6	shell membrane	0.3	80.0	$1\frac{1}{4}$
Shell gland	10.1	calciferous shell	6.1	98.4	20
Cloaca	6.9	mucus	0.1		$\frac{1}{60}$

   a. How long does it take to make an egg?

   b. How long is the oviduct?

   c. What is the weight of the egg solids?

   d. On average, how much albumin is added per centimeter length of magnum cells?

   e. On average, how much albumin is made per hour?

   f. What is the rate of calcium salt deposition in grams per hour?

   g. What assumptions did you make?

18-11   An ant can lift 50 times its own weight. A bee can pull 300 times its weight. Fleas can jump 200 times their own length. One caterpillar (*Polyphemus* moth) consumes 86,000 times its birth weight in 24 days.

   The average biology student is 6 feet tall and weighs 150 pounds. The average human baby is 21 inches long and weighs 7 pounds.

   If humans could perform proportionally to these insects,

   a. How much could a biology student lift?

   b. How much could a biology student pull?

   c. How far could a biology student jump?

   d. How many pints of milk would a baby average per day?

   e. What assumptions did you make?

18-12   A hibernating animal may awaken at intervals and return to the hibernating state. Arousal from hibernation accelerates metabolism, often with an overshoot during the process.

   One example is the hamster, which when hibernating consumes oxygen at the rate of 0.5 ml/kg/hr. This rate of oxygen use is directly proportional to the energy expended by the hamster. During arousal, O_2 consumption rises to 8000 ml/kg/hr and then achieves a stable rate when alert of 5000 ml/kg/hr. The arousal may take 3 hours. The hamster used in this experiment weighed 50 grams.

Chapter 18

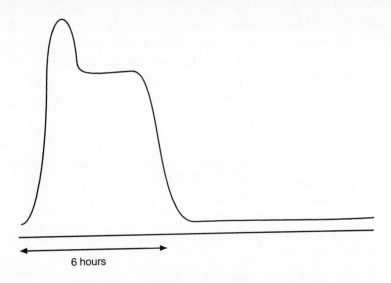

Arousal Period for Hibernating Hamster

a. What percentage is the hamster's metabolism when hibernating compared to its metabolism during its normal alert state?

b. What is the percentage of overshoot as the hamster is arousing?

c. What percent of the energy is expended by the hamster if it hibernates, waking once every 4 days, as compared to being alert the entire time?

d. What assumptions did you make?

18-13 Cows have four stomachs (or a four-compartmented stomach), one of which is called the *rumen*. The rumen may be as large as 1/7 of the animal. In the rumen, an anaerobic culture exists that contains 10^{11} bacteria and 10^6 protozoa per milliliter. Essentially, ruminants use the microorganisms as their food source, either by absorbing the fermentation products or by digesting the microorganisms themselves.

An estimate of 10^{12} bacteria or protozoa weigh a gram. A 500-pound steer can gain weight at 2 pounds per day.

a. What is the weight in pounds of microorganisms in the rumen?

b. If the rumen contents were emptied and refilled in 24 hours, what is the efficiency of biomass conversion from microorganisms to steer?

c. What assumptions did you make?

Chapter 18

18-14 *Circadian rhythms* are daily variations in physiological functions in animals that are related to light-dark cycles. These have been studied most in male lab rats. Below are data for the amounts of two brain chemicals in rats that were exposed to light from 0600 to 1800 hours. (This is using a 24-hour clock, so 0600 corresponds to 6 A.M. and 1800 to 6 P.M.) The measurements were made hourly throughout the day. The amounts are in micrograms of chemical per gram of brain.

Component	Maximum Amount	Time	Minimum Amount	Time
Norepinephrine	0.29	1830	0.22	1230
Serotonin	0.73	1130	0.62	2130

a. How much time elapsed from the time of the maximum to the time of the minimum amount of each chemical?

b. What percentage larger is each maximum than the corresponding minimum?

c. Drugs exist that can raise norepinephrine and serotonin concentrations for pain control. If drugs are used to enhance each chemical, when should they be administered?

d. What assumptions did you make?

18-15 Development rate may be related to environmental temperature. The following table gives the time in hours for hatching of house fly eggs at temperatures from 15° to 40°C.

Temperature	Time to Hatch
15.0	51.5
17.8	33.3
20.6	23.1
23.3	17.2
26.1	13.5
28.9	10.7
31.7	9.0
34.4	8.1
37.2	7.2
40.0	8.1

a. Plot the data.

b. Does the increased hatching rate versus increased temperature give a linear or nonlinear graph?

c. Does the graph demonstrate an optimum temperature for egg hatching?

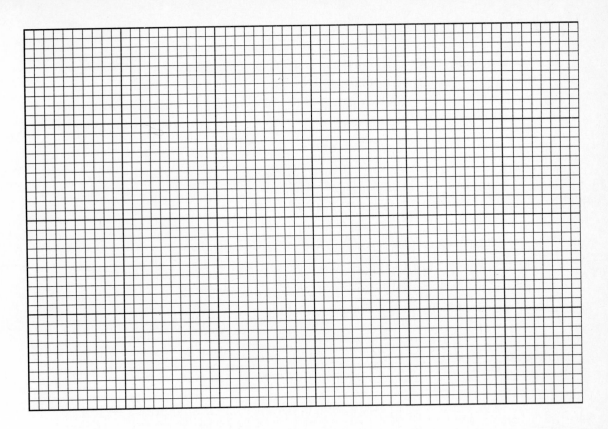

d. Interpolate to estimate the hours to hatch at 25° and at 35°.

e. Compute the rate of decrease of time needed to hatch as the temperature is changed from 15° to 25° and from 25° to 30°.

f. What assumptions did you make?

18-16 The life span of an insect can be modified by the temperature of the environment. Fruit flies, both male and female, live longer in cooler environments, as shown in the following table.

Degrees C	Life span (days)
10	120.5
15	92.4
20	40.2
25	28.5
30	13.6

a. Plot the data.

Because the data are approximately linear, we can find the linear function, $y = mx + b$, whose graph is the straight line that best fits these data. This line is called the *regression line*.

Let (x_i, y_i) represent the ith data point; for example, $x_1 = 10$, $y_1 = 120.5$. If there are n data points, the formulas for the slope m and y-intercept b of the regression line are

$$m = \frac{(\sum x_i)(\sum y_i) - n(\sum x_i y_i)}{(\sum x_i)^2 - n \sum x_i^2}$$

$$b = \frac{1}{n}\left(\sum y_i - m \sum x_i\right)$$

b. Calculate and graph the regression line for the fruit fly data.

c. During which 5-degree temperature interval is the rate of decrease in fruit fly life span closest to the slope of the regression line?

d. What assumptions did you make?

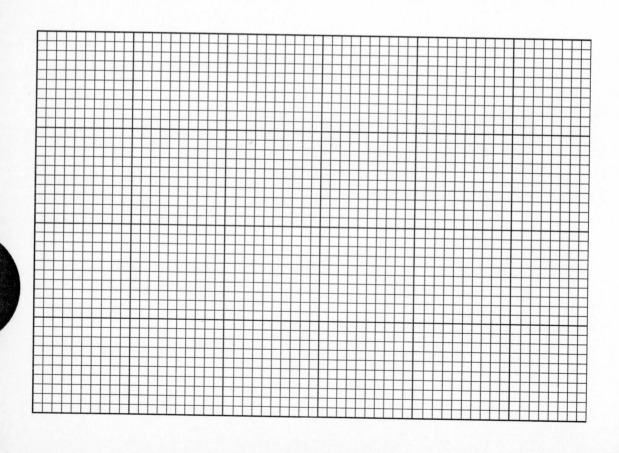

18-17 *Homeotherms* are commonly called warm-blooded animals and *poikilotherms* are called cold-blooded animals. Homeotherms use 98% of assimilated energy in metabolism and "store" 2% as growth. Poikilotherms by comparison use 56% of assimilated energy for metabolism and 44% for growth. For homeotherms, on average, 77.5% of the consumed energy becomes assimilated energy; the efficiency is 77.5%. The corresponding ratio for poikilotherms is 41.9%.

  a. How much food must be consumed by an animal of each type to gain one gram of weight?

  b. What assumptions did you make?

18-18 Measured differences in ionic concentrations across a lobster giant axon in a resting condition are given in the table below. The membrane potential is −70 millivolts.

   For the alga *Nitella translucens*, with a membrane potential −138 millivolts, the corresponding ionic concentrations are also given in the table. All concentrations are given in millimolars.

| | LOBSTER | | ALGA | |
Ion	Outside	Inside	Outside	Inside
Na	465	46	1.0	14
K	10	292	0.1	119
Cl	533	57	1.3	65

To determine diffusion equilibrium, a Nernst equation may be employed.

$$E = 0.058 \log \left(\frac{\text{outside concentration}}{\text{inside concentration}} \right) \quad \text{for cations}$$

and

$$E = 0.058 \log \left(\frac{\text{inside concentration}}{\text{outside concentration}} \right) \quad \text{for anions}$$

where E is the electrical potential in volts.

   Ion pumps usually maintain cells in a nonequilibrium condition. However, when some types of poisons are given to cells or if the temperature is lowered to near freezing, the cells will approach a diffusional equilibrium.

  a. Are any of the ions in lobster nerves or algal cells in diffusional equilibrium?

Chapter 18

b. How far away, as a percentage, do the ion pumps maintain the internal concentrations of cations and anions?

c. What assumptions did you make?

18-19 Cells divide in a rhythmic manner in some tissues. The data in the table below are for three types of tissue: young adult human shoulder skin, rat liver, and rat epidermis. At the listed time of day, the number of cells per 1000 cells that are undergoing mitosis is given.

Time of Day	Shoulder Skin	Rat Liver	Rat Epidermis
4 A.M.	2.05	2.1	8.1
8 A.M.	1.1	13.1	5.75
12 noon	0.9	10.3	6.7
4 P.M.	2.9	4.95	4.9
8 P.M.	1.1	2.7	11.8
12 P.M.	6.05	2.3	14.0

a. Find the daily average number of mitoses of each cell type.

b. Which cell type undergoes mitosis fastest? What percentage faster is this cell type's division than that of the slowest?

c. Within each cell type, how different as a percentage is the highest to the lowest mitosis rate?

d. What assumptions did you make?

18-20 Animals always produce heat because of respiration. The table compares the heat produced by a warm-blooded mouse and a cold-blooded reptile.

MOUSE		REPTILE	
Environmental Temperature	Heat Produced (kcal/mm^2/day)	Environmental Temperature	Heat Produced (kcal/kg/day)
14.6	1741	18	1.2
20.0	1037	26	2.3
24.9	953	32	3.2
29.9	879		
35.3	1009		

The mouse mass-to-surface-area ratio K is 9.0, where

$$K = \frac{\text{area in square centimeters}}{\text{body mass in grams}}$$

a. Convert the mouse data to the units used by the reptile data and plot the data for each.

b. For each animal, do the data indicate a linear or a nonlinear relationship?

c. At what temperature is the most rapid increase in heat production by the mouse?

d. At 26 degrees, which animal generates more heat? By what percentage?

e. On average, how many kcal/kg/day are produced by the reptile for each 1 degree increase in environmental temperature?

f. What assumptions did you make?

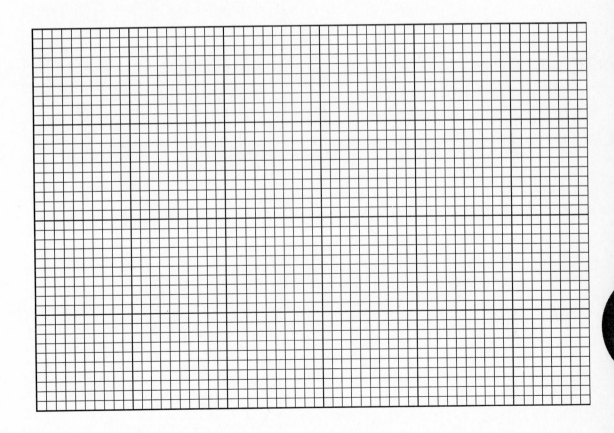

Human Physiology

Answers to Chapter 19 begin on page 269.

19-1   The human ventilation system is a highly branched system of tubes. The lungs branch from 20 to 25 times. The branchings are usually dichotomous; one tube branches into two. The branches terminate in clusters of alveoli that resemble a bunch of grapes. The number of alveoli in a lung varies from 300 million to 500 million.

   a. What is the average number of alveoli in a cluster?

   b. What assumptions did you make?

19-2   When you are at rest, your heart pumps about 5 liters of blood each minute, of which 225 milliliters goes to the heart itself. During stenuous exercise your heart pumps 4 to 6 times more than at rest, and the blood flow to the heart increases from 3 to 4 times.

   a. What is the blood flow to your heart during strenuous exercise?

   b. What percentage of the blood flow goes to your heart when you are resting?

   c. What percentage of the blood flow goes to your heart during strenuous exercise?

   d. What assumptions did you make?

19-3   Both trained, in-shape athletes and nonathletes move about 5 liters of blood per minute when resting. When exercising, athletes move about 30 l/min and nonathletes move about 22 l/min. The heart rate of a resting athlete averages 50 beats/min while that of a nonathlete is 72 beats/min. When exercising, the athlete's heart rate is 185 beats/min while the nonathlete's is 195 beats/min.

   a. What is the stroke volume (milliliters per beat) of a

   • resting athlete?

   • resting nonathlete?

   • exercising athlete?

   • exercising nonathlete?

   b. What are the ratios of the stroke volumes of athletes to nonathletes when

   • resting?

   • exercising?

   c. What assumptions did you make?

19-4 Animals rely on diffusion to obtain nutrients and oxygen and to lose toxins and carbon dioxide. To provide sufficient surface area for diffusion, large animals have lungs. Within the lungs, sacs called *alveoli* facilitate the diffusional gaseous exchange. Each lung contains 300 million alveoli and the average alveolus is 100 to 300 μm in diameter.

   a. What is the range of surface area of one alveolus?

   b. Assuming the average diameter is 200 μm, what is the total surface area of the lungs?

   c. If the skin area of an adult is about 2 m^2, which surface area is larger? By what percentage?

   d. What assumptions did you make?

19-5 The heart moves blood in pulses or beats. At rest, the heart beats about 72 times per minute. The average volume of blood moved per beat is 70 milliliters. The total amount of blood in an adult is 5 liters.

A well-trained marathon runner can move blood 7 times faster than the average person. During exercise, the runner may have a heart rate of 180 to 200 beats per minute.

   a. How much time does it take the average person at rest to move the 5 liters of blood through the heart?

   b. What is the average volume of blood moved per beat by the marathon runner?

   c. Life expectancy includes 65 adult years. If you live them at rest, how many gallons of blood would you pump?

   d. If your heart were used to fill up a 24-can case of 12-ounce cans of your favorite beverage, how long would it take?

   e. The competition pool in the natatorium in Indianapolis used for Olympic trials contains 1 million gallons. How long would it take your heart to fill the pool?

   f. What assumptions did you make?

19-6 An adult pancreas weighs between 50 and 70 grams. The pancreas has both an exocrine and an endocrine function. The endocrine portion is 1% to 2% by weight and consists of about 2 million cells called *islets of Langerhans*. Sixty percent of the islets are involved in insulin production, producing 25 nanograms per kilogram of body weight per minute in fasting conditions.

a. What do the insulin-producing cells weigh?

b. If the average adult weighs 70 kilograms, what is the percentage by weight of the insulin-producing cells?

c. How long will it take for one islet cell to produce enough insulin to equal its own weight?

d. What assumptions did you make?

19-7 The volume of air moved in each breath (the *tidal volume*) under resting conditions is 0.5 liter. About 12 breaths per minute are taken at rest. Under short-term strenuous exercise, the respiratory rate increases 4 times and the tidal volume increases 6 times.

a. What volume of air can be moved per minute at rest and during strenuous exercise?

b. What percentage more air can be moved during strenuous exercise than when at rest?

c. A five-person hot-air balloon contains 140,000 cubic feet of air. How long would it take a person at rest to blow up the balloon?

d. What assumptions did you make?

19-8 Insulin is constantly being made and degraded. A fasting level of insulin is 0.5 nanograms per milliliter of plasma. (After ingestion of glucose it increases to 4 to 5 ng/ml.) The fasting-level secretion of insulin is 25 nanograms per kilogram of body weight per minute. Human insulin has a molecular weight of 5808.

a. How many molecules of insulin are released per minute by a 70 kg adult when fasting?

b. How many molecules of insulin are found in a fasting person's plasma if plasma is 55% of his 5 liters of blood?

c. What assumptions did you make?

19-9 The two kidneys have 2,400,000 nephrons. Nephrons are composed of a glomerulus and a tubule. In the glomerulus the blood is filtered. Proteins and components larger than proteins remain in the blood. Glomerular filtrate (blood plasma and components smaller than proteins) then enters the tubules. The total volume of glomerular filtrate is 125 ml/min. Blood flow to both kidneys is 650 ml/min of the total cardiac output of 5 l/min.

a. What percentage of the blood flows through the kidneys?

b. In the glomerulus, what percentage of the blood enters the tubules?

c. In one day how much blood is "filtered" by the kidneys?

d. Compared to the 70-kg weight of the average biology student, what is the weight of blood filtered per day?

e. What is the volume of the filtrate produced by one glomerulus in one day?

f. What assumptions did you make?

19-10   Animal hormones are produced continuously and are found in low concentration in the plasma. Humans have about 5 liters of blood, 55% of which is plasma. Here is the data for four human hormones:

Gland	Hormone	Molecular Weight	Secretion Rate per Day	Concentration in Plasma
Thyroid	T_3	650	30 μg	120 ng/dl
Thyroid	T_4	777	80 μg	8 μg/dl
Adrenal	cortisol	346	8–25 mg	40–80 ng/ml
Adrenal	aldosterone	360	50–200 μg	0.05–0.2 ng/ml

a. How many molecules of thyroxine (T_3 and T_4), cortisol, and aldosterone are made in a day?

b. How many molecules of thyroxine, cortisol, and aldosterone are found in plasma?

c. What assumptions did you make?

19-11   In an adult, the average length of the small intestine is 2.8 meters and its diameter is 4 centimeters. Because of folds, villi, and microvilli the surface area is 2,000,000 square centimeters. The rate of movement of the contents of the small intestine averages 1 centimeter per minute.

a. What is the surface area of the small intestine in square feet?

b. What percentage increase in surface area is due to the folding and the villi?

c. How long does it take to move food through the small intestine?

d. What assumptions did you make?

19-12 Your tongue contains 10,000 taste buds. They are replaced every 10 days.

a. How many taste buds are made while you are chewing gum during a one-hour exam?

b. What would be the duration of a two-taste-bud kiss?

c. What assumptions did you make?

19-13 A heart "at rest" beats at 72 beats per minute and moves 70 ml of blood per beat. When stimulated by epinephrine (the "fight or flight" hormone), the heart rate increases to double the resting rate and the volume moved per beat triples. Similarly, "at rest" air movement into and out from the lungs is at a rate of 5.6 liters per minute. During maximal exercise, air movement increases to as much as 14 liters per minute due to more rapid and much deeper breathing.

a. What volume of blood is moved per minute by a resting heart?

b. What volume of blood is moved per day by a resting heart?

c. What volume of blood can be moved per minute by a stimulated heart?

d. As a percentage, how much greater is the amount of blood moved by a stimulated heart compared to that moved by a resting heart?

e. As a percentage, how much greater is the amount of air moved during exercise compared to that moved when at rest?

f. What assumptions did you make?

19-14 In adults under normal cool conditions, blood flow to the skin is 250 milliliters per square meter per minute. When environmental conditions change, blood flow to the skin changes. Under cold conditions the flow decreases to 12% of normal, and it can increase during exercise to 750% of normal. Also, during strenuous exercise athletes have recorded sweat loss (almost all water) of 5 to 10 pounds per hour. An adult has 1.7 square meters of skin surface area.

Chapter 19

a. How many milliliters of blood flow through a square meter of skin surface area in an hour during normal, cold, and exercise conditions?

b. How many milliliters of water (sweat) are lost through a square meter of skin surface area during an hour's exercise?

c. What assumptions did you make?

19-15 The skin surface area of an adult can be calculated using the formula

$$\text{skin surface area} = 0.007184 \times \text{weight}^{0.425} \times \text{height}^{0.725}$$

where skin surface area is measured in square meters, weight is in kilograms, and height is in centimeters.

Assume an average adult weighs 150 pounds and is 5 feet 9 inches tall.

a. What is the skin surface area?

Surface areas are important in the functioning of a number of human organs. In an adult, some surface areas are

Small intestine	2,000,000 cm^2
Lung	68 m^2
Sum of glomerular capillaries of the kidney	2 m^2
Sum of liver lobes	400,000 mm^2
Bone cell surface area	1 acre

b. List in order from smallest to largest surface area. Include skin area.

c. What assumptions did you make?

19-16 Up to 1500 ml of water is gained and lost each day by an average student. Assume that 500 ml is lost per day through the lungs. A sedentary individual averages 12 exhalations per minute.

a. How much does the exhaled water weigh in each exhalation?

b. How many water molecules are lost in each exhalation?

The heat of vaporization of water at body temperature is 576 cal/g.

c. How much heat is lost in each exhalation?

d. What assumptions did you make?

19-17 In humans, CO_2 produced by respiration is removed from the cells via the bloodstream and is released into the lungs and ultimately into the atmosphere.

Normal resting breath volume (tidal volume) is 500 ml. Inhaled air is about 0.04% CO_2. Exhaled air is about 5.6% CO_2.

CO_2 is carried in the blood by three pathways:

7% as CO_2 dissolved in plasma

23% as CO_2 attached to hemoglobin

70% as HCO_3^- dissolved in plasma

a. How many CO_2 molecules per exhalation are carried to the lungs by each pathway?

b. What assumptions did you make?

19-18 The respiration of glucose is described by the following equation:

$$C_6H_{12}O_6 + 6\,O_2 + 6\,H_2O \longrightarrow 6\,CO_2 + 12\,H_2O$$

and 686 kilocalories per mole of glucose are released as energy.

a. How many kilocalories of energy are released for each liter of O_2 used?

b. If 1 mole of glucose (about 1/3 pound) is respired, and all of the water vapor generated is exhaled, what would be the volume?

c. If this respiratory water is released as liquid water (urine), what would be the volume?

d. What assumptions did you make?

19-19 For an average biology student who weighs 70 kilograms, each liter of oxygen consumed releases 4.8 usable kilocalories of energy. Each mole of glucose consumed yields 686 kcal of energy.

During an awake but restful period the biology student uses 20 kilocalories per kilogram of body weight per day (basal metabolic rate). During a normally active period the metabolic rate averages 40 kcal/kg body wt/day for a male and 35 kcal/kg body wt/day for a female. During sleep the metabolic rate averages 1.2 kcal/kg body wt/day.

a. How many liters of O_2 is consumed per hour by a student who is resting but awake?

b. If air is 20% oxygen, what is the minimum amount of air moved per hour by a student when resting?

c. If a student were respiring only glucose, how many grams of glucose would be needed per hour when at the basal rate?

d. A female biology student, who is a distance runner, uses 14.3 kilocalories per kilogram of body weight per hour when running. If she practices her running techniques for 2 hours a day in preparation for a race, how many kilocalories would she use per day?

e. What assumptions did you make?

19-20 The blood of a normal adult male contains 5 million red blood cells (RBCs) per microliter of blood. Each deciliter of blood contains 15 grams of hemoglobin. Each RBC contains on average 250 million molecules of hemoglobin. Suppose you gave a pint of blood during a blood drive.

a. How many RBCs did you donate?

b. How many grams of hemoglobin did you donate?

c. How many molecules of hemoglobin did you donate?

d. From the calculations above, what is the approximate molecular weight of hemoglobin?

e. What assumptions did you make?

19-21 At least six different types of blood cells can be detected by using stains or dyes. The most frequent type is the erythrocyte (RBC), of which there are some 5 million per cubic millimeter of blood. Then there are white blood cells, which typically number around 7000 per cubic millimeter. There are many types of white blood cells:

Neutrophils	50–70%
Eosinophils	1–4%
Basophils	0.1%
Monocytes	2–8%
Lymphocytes	20–40%

Helper T cells, at 1000 per cubic millimeter, are the cells whose decrease leads to AIDS symptoms. Helper T cells are in the lymphocytes group above.

a. How many other lymphocytes are present in the blood?

b. What percentage of the blood cells are helper T cells?

c. What assumptions did you make?

19-22 An *action potential* is a nerve signal that lasts 1 to 2 milliseconds. After a signal, a muscle may respond with a twitch that may last for 100 milliseconds. When muscles are stimulated by action potential firings more frequent than every 100 milliseconds, the muscle doesn't have time to fully relax between twitches and it may exhibit tetanic-like contractions, leading to a cramp. In this cramp condition, up to 300 action potential firings per second can occur.

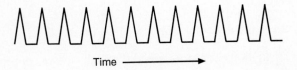

Time ⟶

Action Potentials

a. How many firings per second can occur during one muscle twitch?

b. During a cramp, what is the maximum duration between action potentials?

c. What assumptions did you make?

19-23 Breast-feeding an infant can alter metabolism in the mother by draining several nutrients. Maximum lactation can be 1.5 liters per day.

The chart below gives the components by weight in mother's milk.

Component	Percent by Weight
Water	88.5
Fat	3.3
Lactose	6.8
Casien	0.9
Other proteins	0.4
Ash (calcium and phosphate)	0.2

Chapter 19

a. How many days will it take at maximal lactation for a woman to produce 1 pound of fat in milk?

b. How many grams of calcium and phosphate per day may be lost from the lactating mother's bones?

c. What assumptions did you make?

19-24 The eye constantly makes and loses aqueous humor at the rate of 1 to 2 cubic millimeters per minute. The part of the eye that makes the aqueous humor is called the *ciliary process*, which has an area of 6 square centimeters. The canal of Schlemm allows the loss of aqueous humor. The volume of aqueous humor is 350 microliters. It has a sodium concentration of 163 micromoles per milliliter and a chloride concentration of 134 micromoles per milliliter.

a. How many times is the aqueous humor replaced every day?

b. What volume of aqueous humor is made for each square centimeter of the ciliary process per day?

c. How many micromoles of sodium and chloride pass through the eye each day by this mechanism?

d. What assumptions did you make?

19-25 Each kidney contains 1.2 million nephrons. The nephron length ranges from 2.0 to 4.4 centimeters. The average tubule diameter is 15 micrometers.

a. How many miles of kidney tubule are there in an adult?

b. What is the total kidney tubule surface area?

c. What assumptions did you make?

19-26 The olfactory membrane has an area of 2.4 square centimeters per nostril. The combined olfactory surface area contains 100 million olfactory cells. Each cell has an average of 9 cilia. Each cilium is about 0.3 μm in diameter and 65 μm long.

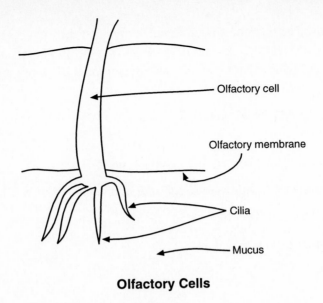

Olfactory Cells

a. What is the total surface area of the cilia?

b. What is the ratio of olfactory cell surface area to the area of the olfactory membrane?

   Natural gas is scented with methyl mercaptan (CH_3SH), which can be smelled when one 25 billionth of a milligram is present in a milliliter of air.

c. If the average inhalation is 500 milliliters, how many molecules of methyl mercaptan must be present for it to be detectable?

d. What assumptions did you make?

19-27 At air temperature $-32.5°C$, the rate of skin cooling is 2000 kilocalories per square meter per hour. It would take only 108 seconds to begin to freeze the skin on your face.

a. If the area of your face is 5 dm^2, what is the average heat loss in calories per minute?

b. How many calories are lost by the time your face begins to freeze?

c. What assumptions did you make?

19-28 Saliva is constantly being released into your mouth. The rate of flow varies among individuals, but an average is about 0.5 milliliters per minute, which goes up to 2.0 ml/min

Chapter 19

when you are eating. The enzyme amylase, which breaks down starch, is a component of saliva. It is released at 100 units per milliliter of saliva. One unit is the amount required to digest in one minute a fixed amount of starch: the amount of starch dissolved at 1% concentration in 5 milliliters of water.

    a. Assume that you eat for a total of 30 minutes each day. What is the volume of saliva produced per day?

    b. How many units of amylase are produced per day?

    c. How much starch by weight can be digested in one minute by one unit of amylase?

    d. How much starch could be digested in one day?

    e. What assumptions did you make?

19-29 HeLa cells are a type of human cancer cell that can be grown in a suspension culture. Their doubling time differs, depending on the temperature. The table below gives the time in hours spent in each of the stages of the cell cycle.

Degrees C	G_1	S	G_2	M	Total
33	26	22.4	12.2	13.0	_____
34	16	14.8	10.5	3.5	_____
36	13	7.4	3.9	1.5	_____
37	10.4	7.0	3.5	0.9	_____
38	7.5	7.6	3.3	0.8	_____
40	15.0	11.2	5.0	2.5	_____

    a. Compute the total time for cell division at each temperature.

    b. What is the time differential in hours between the slowest and the fastest times for cell division?

    c. In which phase, G_1, S, G_2, or M, is the time altered the most when the temperature is adjusted?

    d. At which temperature is the longest percentage of time spent in the interphase?

    e. What assumptions did you make?

Human Development and Reproduction

Answers to Chapter 20 begin on page 280.

20-1 Human semen contains about 100 million sperm per milliliter (ml). About 3 ml of semen is released at copulation. Sperm volume is about 10% of the semen volume.

a. What is the volume of the total number of sperm per ejaculation?

b. What is the volume of one sperm?

c. What percentage of the semen volume is one sperm?

The menstrual flow averages about 50 ml total volume. This flow is generated from the sloughing off of a 3-mm deep lining of the uterus. Assume the uterus is spherical.

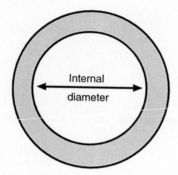

d. What is the calculated internal diameter of the uterus?

e. What is the calculated internal uterine surface area?

f. What assumptions did you make?

20-2 Human sperm can "swim" at a rate of 50 micrometers per second.

a. How long will it take a sperm to swim 1 cm?

Human eggs are fertilized near the ovaries. Fallopian tubes are 10 to 12 cm long. The uterus is about 7 cm long.

b. If sperm are deposited 1 cm from the uterus and eggs are fertilized 1 cm from the ovary, how far did the sperm migrate?

c. If the sperm had to swim the entire distance, how long would it take?

Sperm can survive in a female about 3 days. The egg survives about 10 to 12 hours after its release.

d. How much longer (as a percentage) can a sperm exist compared to a released egg?

e. If sperm deposition occurred at the time of egg release and the sperm had to swim the entire distance to the egg, would the sperm have enough time to swim to the egg?

   A human sperm is 60 μm long and can swim at a rate of 50 μm/sec. An Olympic gold-medal swimmer (assume 6 ft tall) can swim a 50-m distance in 21.91 sec.

f. Compare the speed of the sperm in body lengths per hour to that of the Olympic gold-medal swimmer.

g. What assumptions did you make?

20-3   As boys and girls age, their mean weights increase at different rates. Below are the data in pounds for U.S. boys and girls.

Age	Boys' Weight	Girls' Weight
0	7.8	7.6
1	23.8	21.9
2	29.2	27.6
3	33.5	32.5
4	38.1	37.2
5	42.8	42.3
6	48.2	48.3
7	54.2	54.5
8	61.0	61.9
9	68.4	69.6
10	76.8	78.1
11	85.6	88.4
12	95.2	100.4
13	105.7	110.5
14	119.1	120.1
15	132.3	126.6
16	141.9	130.5
17	147.6	133.5

Answer the following, for both boys and girls:

a. Between which ages is the rate of weight increase most rapid? Between which ages is it the next most rapid?

b. In which year is the percentage of weight gained during the year the largest? The next largest?

c. What assumptions did you make?

Chapter 20

20-4 At birth, human females have about 2 million cells that can produce eggs; at puberty the number is 400,000. A cell that produces an egg is released on average every 28 days. Meiosis may actually be completed after release of the egg-producing cell.

Suppose a woman has a 40-year reproductive lifetime, has regular 28-day cycles, and has two children (gestation period 270 days).

a. What percentage of egg-producing cells has been lost between birth and puberty?

b. How many of the 400,000 potential eggs will be released during the reproductive lifetime of this woman?

c. How many meiosis events per day are needed to produce an egg?

Human males can produce 100 million sperm per day.

d. How many meiosis events per day are needed to produce these sperm?

e. What is the ratio of meiosis events per day in the human male to those in the human female?

f. What assumptions did you make?

20-5 In the developing male child, the testosterone content in the testis is highest at the time of development of the male duct system at gestational week 15. At this time, the amount of testosterone is 2 nanograms per milligram of testis tissue. Testosterone has molecular weight 300.

a. What is the concentration of testosterone in testis tissue?

b. If the testis cells are modeled as spheres of diameter 20 μm, how many molecules of testosterone do these cells contain at week 15?

c. If target cells have 10,000 testosterone receptors and each receptor is bound to testosterone, how many cells could be stimulated by 1 milligram of fetal testis tissue?

d. What assumptions did you make?

PART VII ECOLOGY

CHAPTER 21

Community Ecosystems and Ecology

Answers to Chapter 21 begin on page 284.

21-1 Weather on this planet is due in part to wind currents, and wind currents are modified by frictional interaction with the earth's surface. Because of the rotation of the earth, the surface at the equator moves 425 meters per second. The speed of the earth's surface rotation is less at higher latitudes and is zero at the poles. The radius of the earth is 3964 miles.

a. What is the speed of the earth's rotation at latitudes 20°, 40°, and 60°?

b. What assumptions did you make?

21-2 A Finnish falcon during its nesting period eats voles. Both the male and female of the nesting pair bring voles back to their young for consumption. At two weeks post hatch, the days are about 15 hours long. The male falcons average 44.8% of that time hunting and the females 9.5%. The delivery rate of prey per hour is 0.46 for the males (to the female and the fledglings) and 0.19 for the females (for the fledglings). The average vole size is 20.1 grams. The average number of fledglings is 4.6 per nest. Body mass of an adult male falcon averages 181 grams and the body mass of the female 222 grams.

a. How many total prey were delivered per nest per day?

b. How many grams of food were delivered per day?

c. What percentage of the body mass of the male did he gather per day in prey?

d. What assumptions did you make?

21-3 An albatross can stay at sea for long periods of time, drinking seawater and eating fish and squid. Albatrosses are most common in the Southern Hemisphere between 40° and 50° latitude. An albatross can circle the earth in a year, "riding the wind." At the equator, the circumference of the earth is 24,906 miles.

a. How far can an albatross fly in a year?

b. How far can an albatross fly in a day?

c. What assumptions did you make?

21-4 Brown-headed cowbirds originally lived on the northern Great Plains and ate insects kicked up by bison. Because they were always moving around, they developed the ability to lay their eggs in other birds' nests. As forests have been cleared for farming, there has been an explosion in the population of cowbirds. Forest-dwelling species of birds, who have not evolved any means of defense against this parasitism, have been seriously challenged, since a female cowbird can lay up to 50 eggs in one breeding season. It is estimated that 80% of wood thrush and scarlet tanager nests are parasitized.

It is estimated that a cowbird will not go farther than 5 miles into a forest to find a nest to parasitize, so that the interiors of large, contiguous areas of forest are still safe.

Suppose that a forest is square, with side length 50 miles.

a. What percentage of the forest area is safe nesting area?

To explore the impact of road building on the ecosystem, suppose a road, in a clearing 100 feet wide, is built right through the center of the forest.

b. What percentage of the forest was lost?

c. What percentage of the safe nesting area was lost?

d. What assumptions did you make?

21-5 Marine iguanas on two Galapagos islands eat algae at low tides. They were observed and their stomach contents flushed and dried and the caloric content determined. When average 550 gram iguanas were compared at these two sites, the following data were collected.

	Site A	Site B
Feeding time (min/day)	23.2	81.7
Dry mass eaten (g/day)	2.84	1.72
Number of bites/day	682	3041
Energy (kJ/g)	13.8	10

a. Which site yields more energy per day, and how many times more?

b. Which site yields more energy per bite, and how many times more?

c. Which site yields more energy per minute feeding time, and how many times more?

d. What assumptions did you make?

21-6 Estimates of the amount of pasture available for livestock can be made by satellite sensors measuring red and infrared light. These sensors measure the greenness of plant canopies. Argentinian pastures were measured and an estimate of the number of livestock that could be fed on the available land area (the stocking rate) was determined. In these measurements, the stocking rate varied from 5 kilograms per hectare to 302 kg/ha.

    a. What is the range of land area (in acres) required to support a 500-pound steer on these pastures?

    b. What assumptions did you make?

21-7 Some soils lack micronutrients, and therefore the primary production of the area is equivalent to a desert or the open ocean. Two examples of micronutrient deserts exist in southern Australia. These deserts can support 40 times the number of sheep if zinc is added in one case and copper in the other. In both cases, 7 pounds of sulfate salts per acre was all that was needed. In other depleted soils, 1 ounce per acre of molybdenum in the form of molybdic acid is enough to last 10 years.

    a. How many molecules of $ZnSO_4 \cdot 7H_2O$ are needed per square foot of soil surface?

    b. How many molecules of $CuSO_4 \cdot 5H_2O$ are needed per square foot of soil surface?

    c. How many molecules of $Na_2MoO_4 \cdot 2H_2O$ are needed per square foot of soil surface?

    d. What assumptions did you make?

21-8 Australian water pythons eat dusky rats and not much else. In order to approximate the size of the dusky rat population, ecologists set traps along a 500-m transect, a straight pathway through the area. Rat populations varied from 20 to 140. The number varies mostly because of the amount of rain. Python reproduction is linearly related to the number of rats in the preceding wet season—more rats, more pregnant pythons. However, the number of fertile eggs per nest is about the same at 15 eggs per nest. Also, the size of the eggs is about the same at 55 grams.

At 55 rats per 500-m transect, 40% of the female pythons are pregnant, while at 110 rats per 500-m transect, 90% of the females are pregnant.

    a. Find the equation that relates the percentage of pregnant pythons to the number of rats and graph it.

    b. What percentage of pythons would be pregnant if there were 20 rats the previous year?

    c. According to this assumption of linearity, how small would the rat population have to be to reduce the percentage of pregnant pythons to zero?

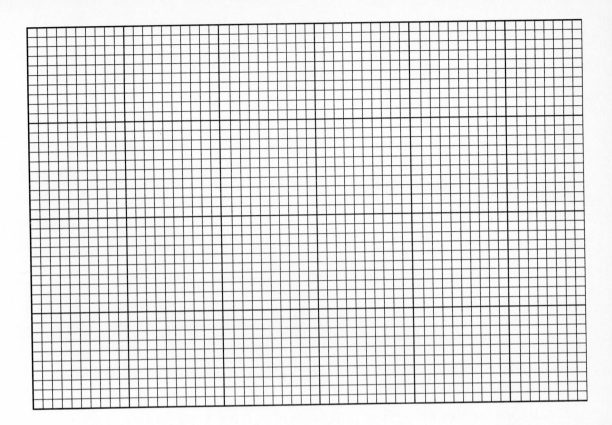

d. Does the rat population ever get large enough that all the pythons could be pregnant?

e. How many rats are needed to increase the number of pregnant pythons by 10%?

f. If the average nonpregnant python weighs 5 kg, how much does one weigh about to oviposit?

g. What assumptions did you make?

21-9 Plants use phosphorus and potassium from soils composed of particles that were originally rocks. The types of rocks in soil vary, so that P and K levels vary also. Topsoil from Nigeria has 200 kg/ha total P and 700 kg/ha total K. Topsoil from Wisconsin has 1936 kg/ha P and 38,180 kg/ha K. Up to 1% of total P and K is available to plants within a growing season.

a. What amounts of P and K are available in pounds per acre in Nigerian soil and in Wisconsin soil?

b. Wheat contains 12 kg/ha P and 15 kg/ha K. Does each soil have enough P and K to generate a wheat crop?

c. What assumptions did you make?

21-10 Soil fertility can be sustained by the addition of elements such as calcium and magnesium (as agricultural lime) and nitrogen, phosphorus, and potassium (from either organic or inorganic amendments). A corn crop (*Zea mays*) can yield about 9500 kg/ha of grain. At this grain yield, the corn stalk mass is around 10,200 kg/ha. This biomass contains (in kg/ha)

	N	P	K
Grain	151	26	37
Stalks	112	18	135

Commercial fertilizer is commonly added to obtain maximum yield. If fertilizer was added at the rate of 200 kg/ha of N, 50 kg/ha of P, and 100 kg/ha of K,

a. Did the soil have a net gain or loss of each of N, P, and K this crop year if the stalks were removed?

b. Did the soil have a net gain or loss of each of N, P, and K this crop year if the stalks were not removed?

c. What assumptions did you make?

21-11 An estimated 3.59×10^{17} gallons of water exists on this planet. About 0.75% of this is freshwater in rivers, lakes, and groundwater. There is about 60 times as much groundwater as water in rivers and lakes. The atmosphere also contains about 0.001% of the total water. The United States averages 4300 billion gallons of rain per day.

a. How much water is in rivers and lakes?

b. How much water is in groundwater?

c. How much water is in the atmosphere?

d. Arrange in order: river and lake water, groundwater, atmospheric water, and U.S. daily rainfall.

e. What assumptions did you make?

21-12 One estimate of the volume of glacier ice is 24,000,000 km^3. The area of the ocean is 361 × 10^6 km^2.

    a. If a dyke were erected along the coastline and all of the glacier ice melted, how many feet high would the dyke need to be to prevent flooding?

    b. What assumptions did you make?

21-13 The amount of CO_2 in the atmosphere as measured in Hawaii has increased from around 320 ppm by volume in 1960 to about 360 ppm in 1995. One thing environmentalists wish to do is to predict what will happen in the future. One mechanism is to extrapolate from existing data.

    a. If the amount of CO_2 is increasing linearly, what will be the concentration in 2030?

    b. If the amount of CO_2 is increasing exponentially, what will be the concentration in 2030?

    c. What assumptions did you make?

21-14 The amount of nitrogen released into one Cape Cod bay has been measured. Nitrogen input to the watershed is 115,000 kg annually. It is due to three primary sources: the atmosphere 56%, fertilizer use 14%, and wastewater, primarily domestic septic systems, 27%. Of these amounts, 89% of the atmospheric nitrogen is retained by the watershed, 79% of that from fertilizer is retained, and 65% of that from wastewater is retained.

    a. How much nitrogen enters the bay annually?

    b. What percentage of the nitrogen load to the bay is due to each of the atmosphere, fertilizer use, and wastewater?

    c. What percentage of the nitrogen added annually to the watershed ends up in the bay?

    d. What assumptions did you make?

21-15 Approximately 320 million acres were harvested in the United States both in 1910 and in 1990. During that time, the U.S. population increased from around 90 million to about 250 million.

Chapter 21

In 1910, about 290 million acres were harvested for domestic consumption. In 1990, about 220 million acres were used for domestic consumption.

a. How many acres per person were used for domestic consumption in 1910?

b. How many acres per person were used for domestic consumption in 1990?

c. How many more people could 100 acres of land support in 1990 than in 1910?

d. How many times more land was farmed for export in 1990 than in 1910?

e. What assumptions did you make?

21-16 The flow of mass from one trophic level to another is usually said to be limited by the ten percent law. This law works for populations, but when employed on an individual basis is not especially meaningful.

A mature vegetarian consumed 2 pounds of food daily. At the end of one year she weighed 1/2 pound more.

a. What was the percentage of biomass conversion at this trophic level?

b. If she had indeed gained 10%, how much more would she weigh?

Assume that the vegetarian was 3 months pregnant at the beginning of the year, had a 7-pound baby and lost an additional 20 pounds right after the birth. She nursed the baby for 6 months. At the end of the year, she had gained the 1/2 pound relative to the start of the year and the baby weighed 12 pounds.

c. What was the percentage of biomass conversion at this trophic level?

d. What assumptions did you make?

21-17 Eating habits have changed with advances in food processing and transportation. This processing and transportation allow for the potential of bacterial contamination. The Public Health Laboratory Service in England has suggested guidelines for some ready-to-eat foods. The process is to collect 1 gram of the food and determine how many colony-forming units (cfu) will develop on agar plates in the presence of O_2.

Food	Serving Size	Satisfactory (cfu/g)	Unsatisfactory (cfu/g)
Confectioneries	75 g	$< 10^3$	$> 10^5$
Cooked meats	60 g	$< 10^4$	$> 10^6$
Sandwiches and salads	100 g	$< 10^5$	$> 10^8$

One gram of fertile soil may contain 2.5 million bacteria.

a. What is the maximum number of cfu in a satisfactory serving of each of the products?

b. How many times as many cfu are required to make each satisfactory type unsatisfactory?

c. How many grams of each of the three satisfactory foods listed would be required to yield as many bacteria as are in 1 gram of soil?

d. What assumptions did you make?

21-18 Ecologists often try to compare two or more communities by comparing their diversities. In the table below, data are given for six tropical forests. The number of species of birds in each family A_1 thru A_{11} in each forest is given in the table. The families are grouped according to their primary food source.

	Costa Rica 1	Costa Rica 2	Panama 3	Malaysia 4	Gabon 5	Liberia 6
A_1	0	0	1	0	0	0
A_2	3	3	4	0	1	2
A_3	0	1	0	0	1	1
A_4	4	4	3	1	1	0
A_5	2	4	1	2	2	1
A_6	3	3	2	2	1	0
A_7	5	6	3	1	4	5
A_8	4	5	12	15	11	12
A_9	2	1	4	5	3	3
A_{10}	2	2	4	0	0	0
A_{11}	1	1	0	1	1	2

Many different formulas have been suggested to calculate a diversity index. One of the most common is the Shannon index H:

$$H = -\sum \left(\frac{n_i}{N} \times \log \frac{n_i}{N} \right)$$

where

$\sum$ indicates the expressions inside the parentheses are to be calculated for each of the 11 families and then summed,

n_i is the number of organisms in family A_i,

N is the total number of organisms in the community.

a. Using this formula and these data, compute the diversity level of each tropical forest and rank them in order.

b. What assumptions did you make?

21-19 The customary rate for deep tillage done in the fall is $15 an acre. This process, if done every other year, will increase the yield. For soybeans, the average yield per acre is 35 bushels, which is increased from 1.6 to 3 bushels with tillage. For corn, the average yield per acre is 110 bushels, increased from 2.5 to 5 bushels with tillage. (These numbers hold for clayey soils.)

a. What price per bushel of soybeans would be required to break even and pay for the cost of tillage?

b. What price per bushel of corn would be required to break even and pay for the cost of tillage?

c. If the selling price for soybeans turns out to be $5.50 during these two years, what increase in the yield is needed to break even?

d. If the selling price for corn turns out to be $1.95 during these two years, what increase in the yield is needed to break even?

e. What price per bushel of soybeans would be required to break even and pay for the cost of tillage if the money to pay for it must be borrowed for one year at 10% simple interest?

f. What assumptions did you make?

21-20 World ecosystem net primary productivity is estimated to be 162.4×10^9 metric tons (t) per year. About two-thirds of this productivity is land and freshwater based, and the other third is marine based (shelf and open ocean).

The tropical rain forest area is 17×10^6 square kilometers, and it accounts for about one-third of land-based productivity. The open ocean area is 332×10^6 square kilometers and its rate of production is 127 grams per square meter per year.

a. What is the productivity of tropical rain forest in grams per square meter per year?

b. How much of the marine productivity comes from the open ocean?

c. How many times more productive is the tropical rain forest than the open ocean?

d. What assumptions did you make?

21-21 A tropical rain forest in Puerto Rico has incoming solar radiation 3830 kilocalories per square meter per day. The gross production of the forest is 131 kcal/m^2/day. The net production is 15.2 kcal/m^2/day. The amount stored in wood is 0.72 kcal/m^2/day.

Nontropical forests store about the same amount of energy in wood as tropical forests, even though the tropical forest gross productivity is 1.7 times that of temperate forests and 2.7 times that of boreal forests.

a. How much energy is stored in wood in kilocalories per square meter per year?

b. What percentage of the gross production is stored in wood in the tropical, temperate, and boreal forests?

c. What assumptions did you make?

21-22 Gasoline and diesel engines are commonly used for transportation in the United States. The U.S. Environmental Protection Agency measured autos and trucks in 1970 and found the following average release of air pollutants:

Pollutant	Amt. from Gasoline (g/mi)	Amt. from Diesel (lb/1000 gal)
Carbon monoxide	80	225
Nitrogen oxides	4.6	370
Hydrocarbons	12	37
Aldehydes	0.36	3
Particulates	0.6	13
Organic acids	0.13	3

Assume that gasoline delivers on average 25 miles per gallon and diesel fuel 20 miles per gallon.

a. Which fuel generates the most total pollutants per mile?

b. How many miles per gallon would you need to get from diesel fuel in order to produce the same amount of nitrogen oxide that is produced by gasoline?

c. Which fuel is cleaner with respect to particulates? How many times cleaner is it?

d. What assumptions did you make?

21-23 The size of a group of muskoxen increases linearly with increasing wolf density. Muskox group size also is different in summer and winter.

After collecting much data, the statistical line of best fit (the *regression line*) for the data was computed to be

$$y = 0.063x + 16.76$$

in winter and

$$y = 0.0546x + 7.76$$

in summer. Here y is the muskox group size and x is number of wolf sightings per 100 hours. Plot these two lines and estimate

a. the number of muskox in a group at 100 wolf sightings in 100 hours in the summer,

b. the number of wolf sightings estimated if the muskox group is 30 in the winter.

c. How much larger is the muskox group size if 200 wolves are sighted per 100 hours in the winter as compared to 200 wolves sighted per 100 hours in the summer?

d. What assumptions did you make?

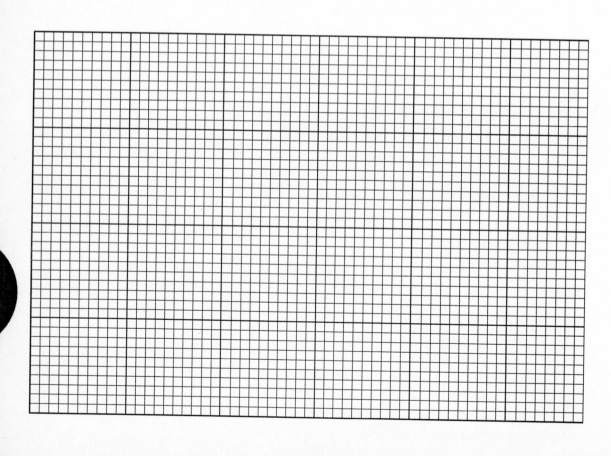

21-24 You are a seed grower of red clover. You know that your seed set is related to bee activity. As a sideline, you sell honey. Of course you wish to maximize your income. You set up trial plots on 1 acre of land and obtain the results in the table. The approximate number of seeds per pound of red clover is 275,000.

Colonies of Bees per Acre	Seeds per Square Yard	Pounds of Honey per Acre per Colony
0	35	0
1	45	103
3	72	92
5	117	73

a. What was the percentage increase in red clover seed with 5 bee colonies compared to 1 bee colony?

b. What was the percentage increase in honey with 5 bee colonies compared to 1?

c. If honey sells for $2.65 per pound and clover seed sells for $1.50 per pound, compute the income per acre in each case.

d. If you decide to purchase 5 bee colonies, what is the maximum you could afford to pay for each of them and still get a 10% return on your investment?

e. What assumptions did you make?

21-25 A female moth ready to mate releases a chemical called a *pheromone*. The pheromone attracts male moths. An amount as small as 0.01 μg is released, which is then dispersed by the wind over a very large volume of air, yet can still be effective 4 km away.

Suppose 0.01 μg is released and the wind spreads it over half of a cone, 4 kilometers long and 200 meters in diameter at its end; see the figure. The molecular weight of the pheromone is 282.

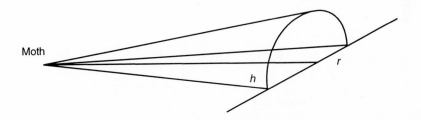

Wind Dispersal of a Pheronome

a. What is the average weight per cubic meter of the pheromone in this conical volume?

b. How many molecules are dispersed in this volume?

c. How many molecules are there per square meter in this volume?

d. What assumptions did you make?

Population Ecology

Answers to Chapter 22 begin on page 294.

22-1 When one cannot collect or count all of the animals in an area, a capture-recapture sampling method can be employed. Let N be the size of the population we are trying to estimate. Suppose during a first capture, N_1 animals are captured, marked, and released. Then some time later during a second capture, suppose N_2 animals are captured, of which M of them are marked from the first capture. Assuming that the ratios $M:N_2$ and $N_1:N$ are equal, we can solve

$$\frac{M}{N_2} = \frac{N_1}{N}$$

to obtain

$$N = \frac{N_1 N_2}{M}$$

In a meadow, 100 baited traps were randomly spaced 20 to 30 meters apart. Twenty-three field mice (voles) were trapped and 22 were marked and released (one died during the trapping and marking process). One week later the traps were set again. Thirty-one voles were trapped this time, with 11 of them marked.

a. How many voles are estimated to be in this population?

b. If the total area was 10 hectares, what is the density of voles in the meadow?

c. What assumptions did you make?

22-2 Basidiomycota produce many spores. One estimate is that one basidiocarp with diameter 7.5 cm, stem diameter 2 cm, and gill surface area 200 cm^2 produces 40 million spores per hour.

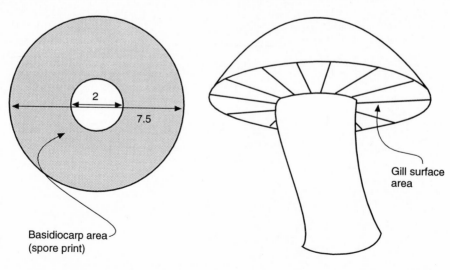

Basidiocarp area
(spore print)

Gill surface
area

Spores are football-shaped with length and width diameters 0.75 μm and 0.375 μm, respectively. They can be approximated by an ellipsoid, which has volume

$$V = \frac{4}{3}\pi a^2 b$$

where a is the shorter semidiameter and b is the longer semidiameter.

a. How many spores are produced in 1 cm^2 basidiocarp area in one minute?

b. How many spores are produced in 1 cm^2 gill surface area in one minute?

c. Using the basidiocarp you make a spore print. After 4 hours, how many spores would you have collected?

d. What is the total volume of the spores from this print?

e. What assumptions did you make?

22-3 Below are human population data for 1995.

Region	Pop. Size (millions)	Birth Rate (per thous.)	Births (millions)	Death Rate (per thous.)	Deaths (millions)	Growth Rate	Doubling Time (years)
Africa	728	42	_____	14	_____	_____	_____
Asia	3458	25	_____	8	_____	_____	_____
Europe	727	12	_____	11	_____	_____	_____
Lat. Amer.	482	26	_____	7	_____	_____	_____
N. Amer.	293	16	_____	9	_____	_____	_____
Oceania	28	19	_____	8	_____	_____	_____
World	_____	_____	_____	_____	_____	_____	_____

The growth rate of a region can be calculated by taking the number of people born that year minus the number who died that year and dividing by the population size. The doubling time t_2 (the number of years required for the population to double) can then be calculated from

$$t_2 = \frac{\ln 2}{\text{growth rate}}$$

a. Calculate the total population of the world.

b. Calculate the number of births and deaths for each of the regions and the world.

c. Calculate the birth rate and the death rate for the world.

Chapter 22

d. Calculate the growth rate and doubling time for each region and the world.

e. What assumptions did you make?

22-4 Below is a chart of estimated world human population over the last 10,000 years, since the beginning of agriculture.

Year	Population Size (in millions)	Growth rate
8000 B.C.	5	
4000 B.C.	86	0.0007
A.D. 1	133	_____
1650	545	_____
1750	728	_____
1800	906	_____
1850	1130	_____
1900	1610	_____
1950	2520	_____
1960	3021	_____
1970	3697	_____
1980	4444	_____
1990	5285	_____

During many periods, the population appears to grow exponentially, following the law

$$P_1 = P_0 e^{rT}$$

where

P_0 is the population at the beginning of the period,

P_1 is the population at the end of the period,

T is the length of the period,

r is the exponential growth rate.

Knowing P_0, P_1, and T we can compute r using the formula

$$r = \frac{\ln(P_1/P_0)}{T}$$

For example, during the first period from 8000 B.C. to 4000 B.C., $r = \ln(86/5)/4000 = 0.0007$.

a. Plot the data.

b. Compute the rest of the exponential growth rates and plot them.

c. If the exponential growth rate stays steady at its 1990 value, what will be the world population in 2040, about the time you retire?

d. What assumptions did you make?

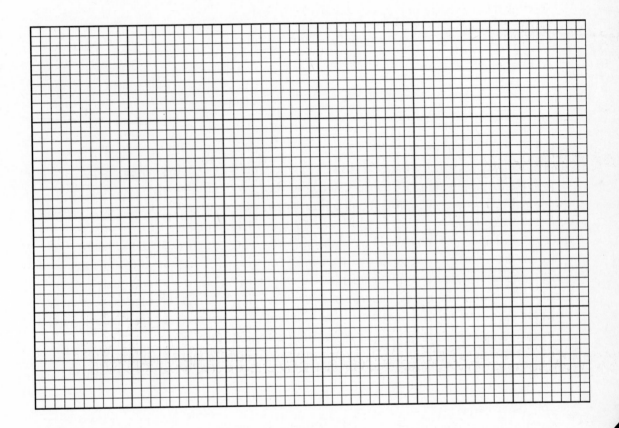

22-5 Populations that grow exponentially follow the law

$$N_t = N_0 e^{rt}$$

where t is the time, measured in this case in days, N_t is the population after t days, N_0 is the initial population, e is the base for the natural logarithms, and r is the growth rate. Observed values of r are

Chapter 22

0.0003 for humans,

0.015 for rats,

60 for *E. coli.*

Assume you start with a male and a female of each sexual species at time $t = 0$.

a. How long will it take for two humans to create one more?

b. How many rats are in the population after 3 years?

c. How many bacteria have accumulated in one day?

d. How much time is required to have as many rats as *E. coli* has in one day?

e. What assumptions did you make?

22-6 One way to describe the occurrence of an infectious disease in a population of size N is to separate the population into subsets, and describe how their sizes vary with time. For example, if $S(t)$ represents the fraction of the people who are susceptible at time t, then the subset of susceptible people has size $N \cdot S(t)$ at time t. Similarly, we let $E(t)$ represent the fraction of the people who have been exposed, but in whom the disease is still latent, $I(t)$ represent the fraction of the people who are infectious, and $R(t)$ represent the fraction of the people who have recovered and have at least temporary immunity. Then the course of a disease can be described by a sequence of letters such as SEIR, which encodes the progress of a disease that transforms susceptible people to exposed, then infectious, then recovered and permanently immune. Mumps is an example of a disease of this type.

Gonorrhea has a very short latency period and confers no immunity, so is an SIS disease. The complications of gonorrhea among women include pelvic inflammatory disease, which makes them 6 to 10 times more likely to have an ectopic pregnancy, and has a 6% chance of making them sterile.

An important number when studying an SIS disease is the *contact number* σ, the average number of contacts a typical infectious person has during his or her infectious period. When the disease has reached a steady state, $\sigma S = 1$ and the fraction of the population that is susceptible remains constant at $S = 1/\sigma$ and the fraction of the population that is infectious remains constant at $I = 1 - S = 1 - 1/\sigma$.

a. If the steady state value of I is negative or zero, the disease will die out. What values of σ indicate that the disease will die out?

b. During an intensive screening of people who visited public health clinics during 1973–1975, public health officials were able to deduce that the current value of σ for that population had to satisfy the equation

$$1 - \frac{1}{0.9\sigma} = 0.72 \left(1 - \frac{1}{\sigma} \right)$$

What was the value of sigma for this population?

c. What percentage of this population was infectious?

d. What assumptions did you make?

22-7 The HIV virus, which causes AIDS, was first identified in the United States in 1981. By the end of 1991 there were an estimated 200,000 cases, of which 44,000, or 22%, had occured during 1991.

Hemophiliacs lack a blood factor that allows blood to clot in case of an accident. They receive this blood factor as a concentrate that is produced by pooling 5000 to 10,000 units of blood. Before 1985, approximately three-fourths of the 14,000 people with hemophilia contracted AIDS from tainted blood. Since late 1984, the concentrate has been heat-treated to kill the HIV virus.

a. Assuming that in each year prior to 1991 it was also true that 22% of the cases were new that year and 78% had been exposed previously, work backward to compute the number of cases of AIDS at the end of 1984.

b. The population of the United States in 1984 was 235 million. What fraction p of the population had AIDS?

   If the frequency, or probability, of having AIDS is p, the probability of not having it is $q = 1 - p$. In a set of n blood donors, the probability of no one having AIDS is q^n. The probability of at least one having AIDS, and therefore contaminating the blood factor concentrate, is $1 - q^n$.

c. Using the value of p obtained in (b) and $n = 7500$, compute the probability that a unit of blood factor concentrate was contaminated with the AIDS virus.

   Intravenous drug users, who more frequently have AIDS, are also more likely to sell their blood to get money for drugs.

d. If the probability p of a blood donor having AIDS goes up to 0.001, how does this affect the probability of contaminating the blood factor concentrate?

e. What assumptions did you make?

22-8 Populations of species demonstrate various mortality profiles. Often ecologists use a life table to develop survivorship curves. A 1000-cohort sample is typically used.

Here are data from three different species. The second column is the number remaining alive at the beginning of the age period.

Age Period (years)	Mountain Sheep	Squirrels
0–1	1000	1000
1–2	801	796
2–3	789	344
3–4	776	151
4–5	764	54
5–6	734	11
6–7	688	6
7–8	640	0
8–9	571	
9–10	439	
10–11	252	
11–12	96	
12–13	6	
13–14	3	
14	0	

Age Period (months)	Short-Lived Grass
0–3	843
3–6	722
6–9	527
9–12	316
12–15	144
15–18	54
18–21	15
21–24	3
24	0

a. Plot the data with *relative* age on the x-axis and survivors on the y-axis. The domain on the x-axis will then run from 0 to 1.

b. Do sheep, squirrels, and grass display similar survivorship curves?

c. When during the lifetime of each species is the mortality rate highest?

d. At approximately what ages are half of the sheep, squirrels, and grass still alive?

e. What assumptions did you make?

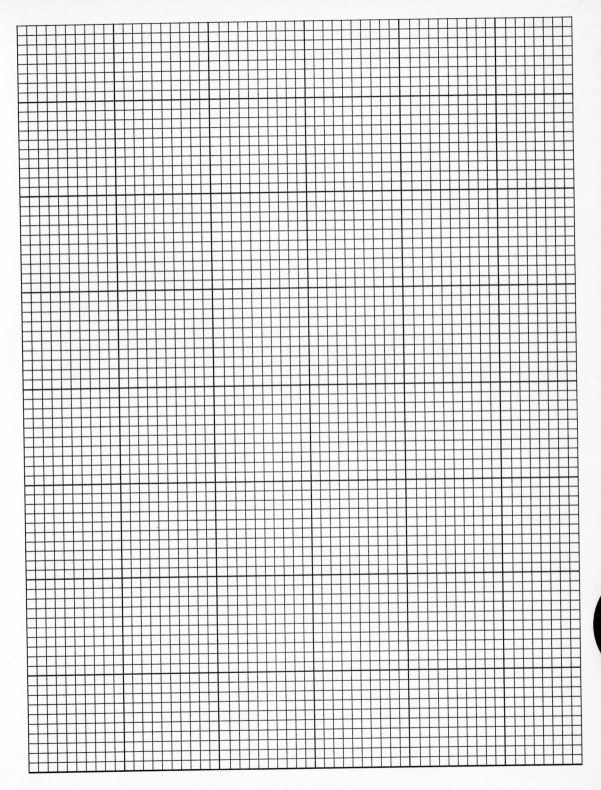

22-9 Poliomyelitis, or infantile paralysis, was a dreaded disease in the United States until the Salk vaccine was developed in 1954 and then the Sabin vaccine in 1957. Striking mostly younger elementary school children, about half the time it left them partially paralyzed. President Franklin D. Roosevelt contracted polio at age 39, was paralyzed from the hips down, and never walked unaided again. Since the early 1970s, the number of cases each year in the United States has averaged about 11, mostly in people who already had impaired immune systems. The data in the table gives the number of cases, paralytic and nonparalytic, in the years 1937–1960. The data are also plotted in the scatterplot. (On the x-axis, year 1 corresponds to 1937, year 2 to 1938, etc.)

Year	Cases	Year	Cases	Year	Cases
1937	9514	1945	13,624	1953	35,592
1938	1705	1946	25,698	1954	38,476
1939	7343	1947	10,827	1955	28,985
1940	9804	1948	27,726	1956	15,140
1941	9086	1949	42,033	1957	5485
1942	4167	1950	33,300	1958	5787
1943	12,450	1951	28,386	1959	8425
1944	19,029	1952	57,879	1960	3190

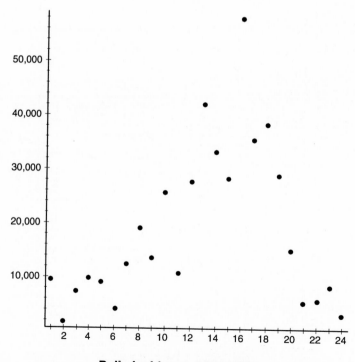

Polio Incidence, 1937–1962

The general apppearance of the graph is an inverted V, with peak about 1952. While the number of polio cases per year from 1937 to 1954 is clearly not a linear function, there is a linear function, $y = mx + b$, whose graph is the straight line that best fits these data. There is another linear function that best fits the data from 1954 to 1960. These lines are called *regression lines*.

Let (x_i, y_i) represent the ith data point; for example, $x_1 = 1$, $y_1 = 9514$. If there are n data points, the formulas for the slope m and y-intercept b of the regression line are

$$m = \frac{(\sum x_i)(\sum y_i) - n(\sum x_i y_i)}{(\sum x_i)^2 - n \sum x_i^2}$$

$$b = \frac{1}{n}\left(\sum y_i - m \sum x_i\right)$$

a. Calculate and graph the two regression lines for the 18-year period 1937–1954 and the 7-year period 1954–1960.

b. If a polio vaccine had never been discovered and the trend of the first regression line had continued, how many cases would have been expected last year?

c. What assumptions did you make?

CHAPTER 1

Chemistry Review

1-1

a.

Element	Number of Atoms		Atomic Weight		Molecular Weight
C	10	×	12.0	=	120
H	12	×	1.0	=	12
N	5	×	14.0	=	70
P	3	×	31.0	=	93
O	13	×	16.0	=	208
Total molecular weight				=	503

b.

Bond	Number	Bond	Number	Bond	Number
C–C	7	H–H	0	N–P	0
C–H	8	H–N	2	N–O	0
C–N	12	H–P	0	P–P	0
C–P	0	H–O	2	P–O	13
C–O	5	N–N	0	O–O	0
P∼O	2				

c. Not found: C–P, H–H, H–P, N–N, N–P, N–O, P–P, O–O

1-2

Bond	LINEAR			RING		
	Number of Bonds	Energy per Bond	Energy in Bonds	Number of Bonds	Energy per Bond	Energy in Bonds
C–C	5	83	415	5	83	415
C–H	7	99	693	7	99	693
C–O	5	84	420	7	84	588
C=O	1	192	192	0	192	0
O–H	5	111	555	5	111	555
Total energy:		Linear	2275		Ring	2251

a. The ring form.

b. The difference in energy between two single bonds and one double bond is 24 kcal/mol. From $24/168 = 0.1428$, we see the energy in the double bond is 14.3% greater.

1-3   **a.** Converting kilograms to pounds,

$$70 \text{ kg} \times \frac{2.205 \text{ lb}}{1 \text{ kg}} = 154.35 \text{ lb, or } 154 \text{ lb}$$

b. We need 18% of 70 kg and of 154.35 lb: 70 kg $\times$ 0.18 = 12.6 kg and 154.35 lb $\times$ 0.18 = 27.8 lb.

c. Similarly, 70 kg $\times$ 0.10 = 7 kg and 154.35 lb $\times$ 0.10 = 15.4 lb.

d. 70 kg $\times$ 0.65 = 45.5 kg and 154.35 lb $\times$ 0.65 = 100 lb.

e. 70 kg $\times$ 0.03 = 2.1 kg and 154.35 lb $\times$ 0.03 = 4.6 lb.

1-4   **a.** The mass of salt in one student is 70 kg $\times$ 0.002 = 0.14 kg. Converting to pounds, 0.14 kg $\times$ 2.205 lb/kg = 0.31 lb.

b.

$$\frac{1 \text{ lb}}{0.31 \text{ lb/student}} = 3.2 \text{ students}$$

So four students are needed.

1-5   **a.** The diameter would be

$$1\frac{7}{16} \times 10^4 = 1.4375 \times 10^4 = 14{,}375 \text{ in.}$$

Dividing by 36 in/yd, we obtain 399 yards. The radius is half this or about 200 yards.

b. The radius is closest to the length of a football field (in fact, two football fields).

1-6   **a.** First we convert 50 mg/dl to g/l:

$$\frac{50 \text{ mg}}{\text{dl}} \times \frac{1 \text{ g}}{10^3 \text{ mg}} \times \frac{10 \text{ dl}}{1 \text{ l}} = \frac{0.5 \text{ g}}{1 \text{ l}}$$

Hence the glucose in 3 liters of blood weighs 1.5 g.
    At 300 mg/dl, 3 liters weighs 6 times as much or 9 g.

b. Glucose weighs 180 grams per mole, thus 1.5 grams is

$$\frac{1.5 \text{ g}}{180 \text{ g/mol}} = 8.3 \times 10^{-3} \text{ mol}$$

Nine grams is 6 times as much or $6 \times 8.3 \times 10^{-3}$ mole $= 0.05$ mole.

c. Using Avogadro's number, 1.5 grams is
$(8.3 \times 10^{-3} \text{ mole}) \times (6.02 \times 10^{23} \text{ molecules/mole}) = 5.0 \times 10^{21}$ molecules and 9 grams is $6 \times (5.0 \times 10^{21})$ molecules $= 3.0 \times 10^{22}$ molecules.

1-7   **a.** Assume that in the human nuclei there is a 1:1:1 ratio of protons, neutrons, and electrons. The positively charged protons and the electrically neutral neutrons have essentially the same mass. So if x represents the mass of the positive particles, we have the equation

$$2x + \frac{1}{2000} x = 70$$

Hence $x = 34.99$ kg. Thus a reasonable estimate is 35 kg.

b. 35 kg.

c. $34.99/2000 = 0.017$ kg.

1-8   **a.** We use the formula for the half-life

$$t_{1/2} = \frac{\ln 2}{r}$$

$$14.3 = \frac{0.693}{r}$$

and obtain $r = 0.0485$.

b. Using the formula for exponential decay from Appendix A,

$$A(t) = A_0 e^{-rt}$$

$$A(7) = 10^8 e^{-(0.0485)(7)} = 7.12 \times 10^7 \text{ cpm/}\mu\text{g}$$

c. Yes, 6 half-lives are $6 \times 14.3 = 85.8$ days.

d.

$$A(365) = 10^8 e^{-(0.0485)(365)} = 2.05 \text{ cpm}/\mu\text{g}$$

1-9   **a.** At concentration 150 mM there are 0.15 moles of KCl per liter, thus 0.15 moles of K^+ and 0.15 moles of Cl^-, for a total of 0.3 moles of ions. Thus they bind 1.2 moles of water.

To find the number of moles of water present in a liter, we use the fact that the atomic weight of water is 18, thus there are 18 grams per mole. Using also the fact that water weighs 1000 grams per liter, we convert to find the concentration of water is

$$\frac{1000 \text{ g/l}}{18 \text{ g/mol}} = 55.56 \text{ mol/l} = 55.56 \text{ M}$$

The ratio of bound water to total water is thus $1.2/55.56 = 0.0216$, so 2.16% of the water is bound.

1-10   **a.** Each such subunit of glycogen requires 11 water molecules to break into 10 molecules of glucose.

b. The molecular weight of the resulting 10 glucose molecules is 1800. Subtracting the molecular weight of 11 water molecules yields the molecular weight $1800 - (11)(18) = 1602$ of the subunit. Thus, the ratio of the weights of the resulting glucose to the original glycogen is $1800/1602 = 1.124$.

If 10% of 1500 g of the liver is glycogen, this 150 g of glycogen yields $150 \times 1.124 = 168.6$ g glucose.

c. Dividing 168.6 by 140, we find there is a supply for 1.2 days or 28.9 hours.

1-11   **a.** The weight of the water is 165 lb $\times$ 0.6 = 99 lb. This must be converted to liters. We use the fact that a liter of water weighs 1 kg.

$$99 \text{ lb} \times \frac{1 \text{ kg}}{2.205 \text{ lb}} \times \frac{1 \text{ l}}{1 \text{ kg}} = 44.9 \text{ l}$$

So a student contains 45 liters of water.

b. The water weighs 44.9 kg = 44,900 g. First we compute the number of moles of water. The molecular weight of water is 18, so it weighs 18 grams per mole.

$$\frac{44{,}900 \text{ g}}{18 \text{ g/mol}} = 2494 \text{ mol}$$

Then using Avogadro's number we compute the number of molecules:
$(2.494 \times 10^3) \times (6.02 \times 10^{23}) = 1.5 \times 10^{27}$ molecules.

c. One-fourth of one-third, in other words 1/12, of the water is plasma. So using the answer to part (a), $44.9/12 = 3.7$ liters of the water is plasma.

1-12 **a.** After the hay is baled, the weight of the dry part is 85% of 85 lb, or 72.25 lb. This dry part is only one-fifth of the same amount of hay before baling. So the total weight when cut of the volume of hay that ends up in one bale is 5 times 72.25 or 361.25 lb. The difference $361.25 - 85 = 276$ lb is the weight of water that evaporated (or was lost) from one bale.

b. From 100 bales, 27,600 lb of water was lost.

c. Converting to kilograms,

$$276 \text{ lb} \times \frac{1 \text{ kg}}{2.205 \text{ lb}} = 125 \text{ kg}$$

d. Converting to liters, using the fact that one liter of water weights one kilogram, 125.3 kg occupies 125.3 l.

e. Converting to quarts,

$$125 \text{ l} \times \frac{1 \text{ qt}}{0.946 \text{ l}} = 132 \text{ qt}$$

f. The weight of water left in one bale is $(0.15)(85) = 12.75$ lb, so 1275 pounds are baled from one acre. Converting to quarts,

$$1275 \text{ lb} \times \frac{1 \text{ kg}}{2.205 \text{ lb}} \times \frac{1 \text{ l}}{1 \text{ kg}} \times \frac{1 \text{ qt}}{0.946 \text{ l}} = 611 \text{ qt}$$

1-13   **a.** ^{12}C has 6 protons and 6 neutrons. ^{14}C has 6 protons and 8 neutrons. Thus it weighs $14/12 = 1.167$ times as much. So the weight of ^{14}C is 116.7% of the weight of ^{12}C; in other words it weighs 16.7% more.

b. The molecular weight of glucose with ^{12}C is 180, while with ^{14}C it is 192. Thus it weighs $192/180 = 1.067$ times as much. So the weight of ^{14}C glucose is 106.7% of the weight of ^{12}C glucose; in other words it weighs 6.7% more.

1-14   **a.** ^{31}P has 15 protons and 16 neutrons. ^{32}P has 15 protons and 17 neutrons. Thus it weighs $32/31 = 1.032$ times as much. So the weight of ^{32}P is 103.2% of the weight of ^{31}P; in other words it weighs 3.2% more.

b. The molecular weight of ATP with ^{31}P is 503, while with ^{32}P it is 506. Thus it weighs $506/503 = 1.00596$ times as much. So the weight of ^{32}P ATP is 100.596% of the weight of ^{31}P ATP; in other words it weighs 0.596% more.

1-15   **a.** The pH 0.84 corresponds to the concentration $10^{-0.84} = 0.145$ M.

b. First we find the number of moles of H^+ in 25 ml by converting

$$0.145\ M = \frac{0.145\ mol}{1\ l} \times \frac{1\ l}{1000\ ml} \times 25\ ml = 3.63 \times 10^{-3}\ mol$$

Next we multiply by Avogadro's number to find the number of molecules:
$(3.63 \times 10^{-3}) \times (6.02 \times 10^{23}) = 2.19 \times 10^{21}$ molecules.

c. The pH of water is 7, so the concentration of hydrogen ions is 10^{-7}M or 10^{-7} moles per liter. Thus there are 2×10^{-7} moles in 2 liters. The number of molecules is $(2 \times 10^{-7}) \times (6.02 \times 10^{23}) = 1.2 \times 10^{17}$ molecules.

d. The total number of molecules of H^+ in 2.025 liters is
$(2.19 \times 10^{21}) + (1.2 \times 10^{17}) = 2.19 \times 10^{21}$ molecules.

e. The total number of moles of H^+ in 2.025 liters is
$(3.63 \times 10^{-3}) + (2 \times 10^{-7}) = (3.63 \times 10^{-3})$ moles. Thus the concentration is

$$\frac{3.63 \times 10^{-3}\ mol}{2.025\ l} = 0.0018\ M$$

f. The pH is $-\log(0.0018) = 2.7$.

CHAPTER 2

Biomolecules

2-1　　**a.** 50.

b. Since amino acid number 1 bonds to number 4, 2 to 5, 3 to 6, ..., 47 to 50, the number of bonds is only 47.

2-2　　**a.** There are more negatively charged molecules at pH 12 when both protons have been lost from many molecules.

b. These are equal because the slopes of the curve at these two points are equal.

c. It is greater at pH 6.02 because the slope of the curve is greater.

d. The highest percentage of zwitterion occurs at pH 6.02 after one proton has been lost from many molecules.

e. They are about the same; in the first case there is an extra proton in many molecules and in the second case one fewer.

f. The buffering capacity is less at pH 6.02 where the rate of change is larger.

g. It is more positive at pH 2.34 where there are more protons than at pH 9.69.

h. It is more positive at pH 6.09 where there are more protons (although fewer than at pH 2.34) than at pH 9.69.

2-3　　**a.** Using

$$pH = pK + \log \frac{c}{u}$$

$$4.0 = 3.65 + \log\left(\frac{c}{u}\right)$$

$$0.35 = \log\left(\frac{c}{u}\right)$$

$$10^{0.35} = 2.24 = \frac{c}{u} = \frac{c}{1-c}$$

$$c = 0.69$$

So 69% of the aspartic acid molecules are charged at pH = 4.0.

b. The remaining 31% are uncharged.

c. Solving the same equation with pH $= 8.5$ yields $c = 0.99998$. So 99.998% are charged.

d. The remaining 0.002% are uncharged.

2-4

a. Adding the molecular weights and dividing by 20, we get 136.75.

b. The molecular weight of lysozyme is

$$(12)(89) + (11)(174) + \cdots + (6)(117) = 16{,}602$$

c. There are 129 amino acids in lysozyme so their average molecular weight is $16{,}602/129 = 128.7$.

d. The amino acids that contain sulfur are methionine and cysteine, but the sulfur in methionine is not free to form bonds. There are eight sulphurs in the cysteine in lysozyme, so four bonds are possible.

e. The basic amino acids arginine, histidine, and lysine have one positive charge, while aspartic acid and glutamic acid have one negative charge. In lysozyme, there are 18 positive charges and 10 negative charges, so the net charge on the enzyme is positive.

2-5 To compute the first entry,

$$pH = pK + \log \frac{c}{u}$$

$$7.2 = 3.87 + \log\left(\frac{c}{u}\right)$$

$$3.33 = \log\left(\frac{c}{u}\right)$$

$$10^{3.33} = 2138 = \frac{c}{u} = \frac{c}{1-c}$$

$$c = .9995$$

So 99.95% of the aspartic acid molecules are charged at pH $= 7.2$.

The remaining computations are done similarly.

Amino Acid in Protein	pK of the Side Chain	% Charged at pH = 7.2	% Charged at pH = 7.4
Aspartic acid	3.87	99.95	99.97
Histidine	6.0	94.1	96.17
Cysteine	8.33	6.90	10.5
Arginine	12.48	5.25×10^{-4}	8.32×10^{-4}

2-6   **a.** For amylopectin, assume the number of 1,4-bonds between branches is 27. Then the polymer has $(1 \times 10^8)/27 = 3.7 \times 10^6$ branches. Each subunit consisting of a linear segment and one branching molecule at the end releases 28 water molecules. So in all, $(3.7 \times 10^6) \times 28 = 1.04 \times 10^8$ water molecules are released.

For amylose, 5×10^5 water molecules are released.

For glycogen, assume the number of 1,4-bonds between branches is 10. Then the polymer has $(3 \times 10^6)/10 = 3.0 \times 10^5$ branches. Each subunit consisting of a linear segment and one branching molecule at the end releases 11 water molecules. So in all, $(3.0 \times 10^5) \times 11 = 3.3 \times 10^6$ water molecules are released.

b. The volume of a plant cell is $(1.8 \times 10^{-3} \text{ cm}) \times (1.8 \times 10^{-3} \text{ cm}) \times (6.0 \times 10^{-3} \text{ cm}) = 1.944 \times 10^{-8} \text{ cm}^3$, or 1.944×10^{-8} ml. Ninety percent of it is water, or 1.75×10^{-8} ml.

An amylopectin molecule releases 1.04×10^8 water molecules. In order to compare this to the volume of water in a plant cell, both must be in the same units. We convert 1.75×10^{-8} ml to molecules. Because one liter of water weighs one kilogram, one milliliter weighs one gram. Now using the fact that the molecular weight of water is 18, and Avogadro's number, we obtain the number of molecules.

$$\frac{1.75 \times 10^{-8} \text{ g}}{18 \text{ g/mole}} \times \frac{6.02 \times 10^{23} \text{ molecules}}{1 \text{ mole}} = 5.85 \times 10^{14} \text{ molecules}$$

The ratio of the amount of water added from an amylopectin molecule to the amount in the cell is $(1.04 \times 10^8)/(5.85 \times 10^{14}) = 1.78 \times 10^{-7}$. So the percentage increase is 1.78×10^{-5} or 0.0000178%.

For amylose, the ratio is $(5 \times 10^5)/(5.85 \times 10^{14}) = 8.55 \times 10^{-10}$. So the percentage increase is 8.55×10^{-8}.

The volume of an animal cell with radius 12.5 μm $= 1.25 \times 10^{-3}$ cm is

$$V = \frac{4}{3}\pi(1.25 \times 10^{-3})^3 = 8.18 \times 10^{-9} \text{ cm}^3.$$

Ninety percent of this is water, or 7.36×10^{-9} ml. As before, this weighs 7.36×10^{-9} g. Finding the number of molecules,

$$\frac{7.36 \times 10^{-9} \text{ g}}{18 \text{ g/mol}} \times \frac{6.02 \times 10^{23} \text{ molecules}}{1 \text{ mol}} = 2.46 \times 10^{14} \text{ molecules.}$$

The ratio is $(3.3 \times 10^6)/(2.46 \times 10^{14}) = 1.34 \times 10^{-8}$. So the percentage increase is 1.34×10^{-6}.

CHAPTER 3

Enzymatics and Energetics

3-1
a. The rate given is per minute, so one-sixtieth of that is 93,300 molecules of product formed per molecule of enzyme per second.

b. Similarly 1150/60 = 19.2 molecules of product formed per molecule of enzyme per second.

c. The ratio is 93,300/19.2 = 4860.

3-2
a. At 36,000,000 substrates per minute, we have 1/36,000,000 minutes per substrate, or

$$\frac{1 \text{ min}}{3.6 \times 10^7 \text{ substrates}} \times \frac{60 \text{ sec}}{1 \text{ min}} = 1.67 \times 10^{-6} \text{sec/substrate} = 1.67 \mu \text{sec/substrate}$$

3-3
a. Substituting in the first equation, we obtain

$$\Delta G = -7.3 \text{ kcal/mol} + 1.98 \times 10^{-3} \text{ kcal/mol} \cdot K \times 298K \times \ln \left(\frac{0.001 \text{ M} \times 0.01 \text{ M}}{0.003 \text{ M}} \right)$$

$$= -10.7 \text{ kcal/mol}$$

b. Similarly,

$$\Delta G = -7.3 \text{ kcal/mol} + 1.98 \times 10^{-3} \text{ kcal/mol} \cdot K \times 298K \times \ln \left(\frac{1.3 \times 10^{-5} \text{ M} \times 0.14 \text{ M}}{0.004 \text{ M}} \right)$$

$$= -11.8 \text{ kcal/mol}$$

3-4
a. Substituting in the first equation with temperature 273 + 20 = 293°C, we obtain

$$\Delta G = -7.3 \text{ kcal/mol} + 1.98 \times 10^{-3} \text{ kcal/mol} \cdot K \times 293K \times \ln \left(\frac{0.001 \text{ M} \times 0.01 \text{ M}}{0.003 \text{ M}} \right)$$

$$= -10.6 \text{ kcal/mol}$$

With temperature $273 + 35 = 318°C$,

$$\Delta G = -7.3 \text{ kcal/mol} + 1.98 \times 10^{-3} \text{ kcal/mol} \cdot K \times 308K \times \ln \left(\frac{0.001 \text{ M} \times 0.01 \text{ M}}{0.003 \text{ M}} \right)$$

$$= -10.8 \text{ kcal/mol}$$

The more negative value of ΔG occurs at the higher temperature.

b. The ratio is $-10.8/(-10.6) = 1.019$, so the increase is 1.9%.

3-5   **a.** The temperature $101°F$ corresponds to $\frac{5}{9}(101 - 32) = 38.33°C$ or to $273 + 33.33 = 311K$.

b. We must compute ΔG in each of the three cases. For the hepatocyte cell, substituting in the first equation, we obtain

$$\Delta G = -7.3 \text{ kcal/mol} + 1.98 \times 10^{-3} \text{ kcal/mol} \cdot K \times 311K$$

$$\times \ln \left(\frac{1.32 \times 10^{-3} \text{ M} \times 4.8 \times 10^{-3} \text{ M}}{3.38 \times 10^{-3} \text{ M}} \right)$$

$$= -11.2 \text{ kcal/mol}$$

Similarly, for the myocyte cell we obtain -11.6 kcal/mol and for the neuron -11.7 kcal/mol. The most negative is the neuron.

c. The energy difference is $11.7 - 11.2 = 0.5$ kcal/mol.

3-6   **a.** The molecular weight of 100 molecules of water is 180. The molecular weight of phosphate, PO_4, is 95. The ratio is $180/95 = 1.9$.

b. Induced fit, because the water loss from the protein would cause a change in shape.

3-7   **a.** The decrease from 18 to 7 kcal/mol is 11 kcal/mol. Dividing by 1.36 we obtain 8.09. Thus, the speed of reaction has been increased by 8.09 factors of 10: $10^{8.09} = 1.23 \times 10^8$.

b. Similarly, $10^{3.68} = 4786$.

c. The ratio is $(1.23 \times 10^8)/(4.79 \times 10^2) = 2.57 \times 10^5$ times faster, or 25,700,000% faster.

3-8

a. At 134,000 grams per mole, we have

$$\frac{1 \text{ mol}}{1.34 \times 10^5 \text{ g}} \times 0.001 \text{ g} = 7.46 \times 10^{-9}$$

moles of enzyme in the milligram. The number of molecules then is $(7.46 \times 10^{-9}) \times (6.02 \times 10^{23}) = 4.49 \times 10^{15}$. As each of these has four active sites, it takes $2 \times 4.49 \times 10^{15} = 8.98 \times 10^{15}$ molecules of the inhibitor to tie up half of them.

3-9

a. The coordinates of P are $(0, 1/V_{max})$. To find the coordinates of Q we let $y = 0$ in the equation $y = mx + b$ and find $x = -b/m$. So the coordinates of Q are $(-1/K_m, 0)$.

b. and c. Draw the graphs with $V_{max} = 2500$ and $K_m = 0.1$ mM.

d. and e. Draw the graphs with $V_{max} = 2400$ and $K_m = 0.075$ mM. The two Michaelis-Menton graphs are

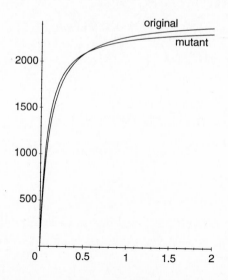

f. The mutant enzyme is better, because with a smaller K_m, the graph of the mutant is above the other in the region of low substrate concentration where the most rapid reaction is taking place.

3-10   **a.** The coordinates of P are $(0, 1/V_{max})$. To find the coordinates of Q we let $y = 0$ in the equation $y = mx + b$ and find $x = -b/m$. So the coordinates of Q are $(-1/K_m, 0)$.

b. For carbonic anhydrase,

$$m = \frac{K_m}{V_{max}} = \frac{9 \times 10^3}{3.6 \times 10^7} = 2.5 \times 10^{-4}$$

For catalase, $m = 0.0045$. For penicillinase, $m = 4.17 \times 10^{-4}$. For chymotrypsin, $m = 0.83$. For lysozyme, $m = 0.2$.

3-11   **a.** The coordinates of P are $(0, 1/V_{max})$. To find the coordinates of Q we let $y = 0$ in the equation $y = mx + b$ and find $x = -b/m$. So the coordinates of Q are $(-1/K_m, 0)$.

d. Letting $[S] = K_m$ and simplifying,

$$\frac{1}{V} = \frac{K_m}{V_{max}} \frac{1}{K_m} + \frac{1}{V_{max}}$$

$$\frac{1}{V} = \frac{1}{V_{max}} + \frac{1}{V_{max}}$$

$$V = \frac{1}{2} V_{max}$$

e. Letting $[S] = \frac{1}{2} K_m$ and simplifying,

$$\frac{1}{V} = \frac{K_m}{V_{max}} \frac{1}{\frac{1}{2} K_m} + \frac{1}{V_{max}}$$

$$\frac{1}{V} = \frac{2}{V_{max}} + \frac{1}{V_{max}}$$

$$V = \frac{1}{3} V_{max}$$

f. Letting $[S] = 4K_m$ and simplifying,

$$\frac{1}{V} = \frac{K_m}{V_{max}} \frac{1}{4K_m} + \frac{1}{V_{max}}$$

$$\frac{1}{V} = \frac{\frac{1}{4}}{V_{max}} + \frac{1}{V_{max}}$$

$$V = \frac{4}{5}V_{max}$$

g.

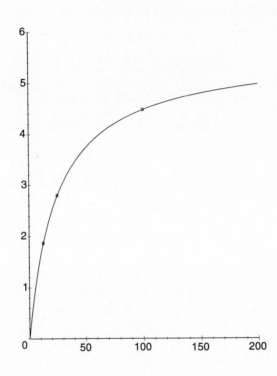

CHAPTER 4

Cell Structure

4-1   **a.** One meter is 10^6 μm. Dividing by 0.6 we find it takes 1.67×10^6 seconds. Converted to more reasonable units,

$$1.67 \times 10^6 \text{ sec} \times \frac{1 \text{ min}}{60 \text{ sec}} \times \frac{1 \text{ hr}}{60 \text{ min}} \times \frac{1 \text{ day}}{24 \text{ hr}} = 19.33 \text{ days}$$

or over 19 days.

4-2   **a.** Assume that the slower-crawling cells move at the average rate $(1 + 0.1)/2 = 0.55$ μm/hr. The faster-crawling cells move at the rate

$$\frac{30\mu\text{m}}{1 \text{ min}} \times \frac{60 \text{ min}}{1 \text{ hr}} = 1800 \text{ } \mu\text{m/hr}$$

The ratio is $1800/0.55 = 3273$, so the infection fighters can crawl 327,000% faster.

4-3   **a.** The microtubule has length 20,000 nm, as does each of its 13 protofilaments, for a total length of 2.6×10^5 nm. Dividing by 8 we find that 32,500 tubulin subunits are required.

4-4   **a.** Each vesicle requires 36 triskelions for a total of $2500 \times 36 = 90,000$ triskelions.

b. 90,000 molecules of ATP.

c. The surface area of one vesicle is $SA = 4\pi (20 \text{ nm})^2 = 5027$ nm^2. So in all, 2500×5027 nm$^2 = 1.26 \times 10^7$ nm^2 or 12.6 μm^2.

4-5   **a.** The circumference of the sphere is $6\pi = 18.85$ μm or 18,850 nm. Dividing by 200 we see it takes 94.25, or about 95 complexes to go once around.

b. Each complex corresponds to one attachment site, so 95 are required.

c. The molecular weight of one tetramere is $2 \times 260 + 2 \times 225 = 970$ kd. Thus, 95 complexes have molecular weight 92,150 kd.

d. $(9.215 \times 10^7)/110 = 8.38 \times 10^5$ amino acids.

4-6   **a.** We compute the number of hydrogen ions inside at each pH and take the difference.

The volume of a lysosome sphere in liters is

$$V = \frac{4}{3}\pi(2.5 \times 10^{-5}\text{cm})^3 = 6.54 \times 10^{-14}\text{ cm}^3 = 6.54 \times 10^{-14}\text{ ml} = 6.54 \times 10^{-17}\text{ l}$$

At pH 5, the concentration is 10^{-5} M $= 1 \times 10^{-5}$ moles per liter. Hence the lysosome contains $(6.54 \times 10^{-17}\text{ l}) \times (1 \times 10^{-5}\text{ mol/l}) = 6.54 \times 10^{-22}$ moles of hydrogen ions. The number of hydrogen ions is therefore $(6.54 \times 10^{-22}) \times (6.02 \times 10^{23}) = 394$.

At pH 7.2, the concentration is $10^{-7.2}$ M $= 6.31 \times 10^{-8}$ moles per liter. Hence the lysosome contains $(6.54 \times 10^{-17}\text{ l}) \times (6.31 \times 10^{-8}\text{ mol/l}) = 4.13 \times 10^{-24}$ moles of hydrogen ions. The number of hydrogen ions is therefore $(4.13 \times 10^{-24}) \times (6.02 \times 10^{23}) = 2.5$.

The difference is 392.

4-7   **a.**

$$V = \frac{4}{3}\pi(0.6\mu\text{m})^3 = 0.905\mu\text{m}^3$$

b. We use the fact that one liter of water weighs one kilogram, so 1 milliliter = 1 cm^3 of water weighs one gram. So water in the same volume as the fat body weighs

$$0.905\ \mu\text{m}^3 \times \frac{1\text{ cm}^3}{10^{12}\mu\text{m}^3} \times \frac{1\text{ g}}{1\text{ cm}^3} = 9.05 \times 10^{-13}\text{ g}$$

c. The fat body weighs 9/10 of this or 8.15×10^{-13} g.

d. The molecular weight is $51(12) + 98(1) + 6(16) = 806$.

e. First find the number of moles:

$$\frac{8.15 \times 10^{-13}\text{ g}}{806\text{ g/mole}} = 1.0 \times 10^{-15}\text{ mol}$$

Then the number of molecules is $(1.0 \times 10^{-15}) \times (6.02 \times 10^{23}) = 602{,}000{,}000$.

4-8

a. Let us assume the top of each microvillus is hemispherical. Then its surface area is $SA = 4\pi r^2/2 = 2\pi(0.05)^2 = 0.0314 \; \mu m^2$. The surface area of the sides of a microvillus is $SA = 2\pi rh = 2\pi(0.05)(1) = 0.3142 \; \mu m^2$. In all, $0.346 \; \mu m^2$.

b. The surface area is $(2 \times 10^9) \times 0.346 \; \mu m^2 = 6.92 \times 10^8 \; \mu m^2 = 6.92 \; cm^2$.

c. The microvilli increase the surface area from $1 \; cm^2$ to about $6.9 \; cm^2$. The increase is 690%.

4-9

a. $(0.1)(0.35) = 0.035$, so 3.5% is protein.

b. Lysine is basic and glutamic acid is acidic. These charges cancel, so the protein portion of the cell wall is neutral. Therefore, the pectins determine the pH.

c. The pectins are acidic, so the cell wall is more acidic.

CHAPTER 5

Membranes

5-1

a. A total of $12 \times 21 = 252$ amino acids are involved in the spanning. The ratio is $252/492 = 0.512$, so 51.2% are involved.

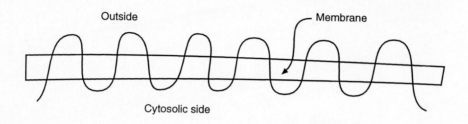

b. Thirteen.

c. There are $492 - 252 = 240$ amino acids divided among the 13 segments. So there are an average $240/13 = 18.5$ amino acids in each.

5-2

a. Twenty hours is 120 ten-minute periods. At 2000 cholesterol molecules each period, 240,000 can be brought into the cell.

5-3

a. In each 50 milliseconds, or 0.05 seconds, 10^9 ions cross. Twenty times as many, or 2×10^{10}, cross in one second. These crossings occur through 1.2×10^6 pores, so $(2 \times 10^{10})/(1.2 \times 10^6) = 16{,}700$ ions cross per protein per second.

5-4

a. An entire turn is a 360° rotation. Each amino acid accounts for 100° so it takes 3.6 amino acids.

b. Each of the 3.6 amino acids contributes a 1.5 angstrom rise, for a total rise of $3.6 \times 1.5 = 5.4$ angstroms or 0.54 nm.

c. It takes $4/0.54 = 7.41$ complete turns.

d. $7.41 \times 3.6 = 26.7$, so it takes at least 27 amino acids.

5-5

a. $SA = 4\pi r^2 = 4\pi (25)^2 = 7854 \text{ nm}^2$.

b. $SA = 4\pi(20)^2 = 5027$ nm^2.

c. At 5×10^6 lipid molecules per square micrometer, there are 5 per square nanometer. So $5 \times 5027 = 25{,}135$.

d. The number of lipid molecules that make up the inner half of the lipid bilayer is $5 \times 7854 = 39{,}270$. They all had to be flipped.

5-6   **a.** 4×10^7.

b. We need to find the difference between the numbers of calcium ions at the two concentrations. The volume of the cell is

$$\frac{4}{3}\pi r^3 = \frac{4}{3}\pi(25 \ \mu m)^3 = 6.54 \times 10^4 \ \mu m^3$$

The number of calcium ions at the higher concentration is

$$\frac{5 \times 10^{-6} \ \text{mol}}{1 \ \text{l}} \times \frac{6.02 \times 10^{23} \ \text{molecules}}{1 \ \text{mol}} \times \frac{1 \ \text{l}}{1000 \ \text{ml}} \times \frac{1 \ \text{ml}}{1 \ \text{cm}^3} \times \frac{1 \ \text{cm}^3}{10^{12} \ \mu m^3} = \frac{3010 \ \text{molecules}}{1 \ \mu m^3}$$

for

$$(6.54 \times 10^4 \ \mu m^3) \times \frac{3010 \ \text{molecules}}{1 \ \mu m^3} = 1.97 \times 10^8$$

molecules in the cell.

At the lower concentration, a similar computation shows there are 3.94×10^6 molecules in the cell. The difference $1.97 \times 10^8 - 3.94 \times 10^6 = 1.93 \times 10^8$ molecules had to be removed. As only 4×10^7 could be removed by calmodulin, the answer is no.

5-7   **a.** We need to find the difference between the numbers of calcium ions at the two concentrations. The volume of the cell is

$$\frac{4}{3}\pi r^3 = \frac{4}{3}\pi(22.5 \ \mu m)^3 = 4.77 \times 10^4 \ \mu m^3$$

The number of calcium ions at the higher concentration is

$$\frac{5 \times 10^{-6} \text{ mol}}{1 \text{ l}} \times \frac{6.02 \times 10^{23} \text{ molecules}}{1 \text{ mol}} \times \frac{1 \text{ l}}{1000 \text{ ml}} \times \frac{1 \text{ ml}}{1 \text{ cm}^3} \times \frac{1 \text{ cm}^3}{10^{12} \text{ } \mu\text{m}^3} = \frac{3010 \text{ molecules}}{1 \text{ } \mu\text{m}^3}$$

for

$$(4.77 \times 10^4 \text{ } \mu\text{m}^3) \times \frac{3010 \text{ molecules}}{1 \text{ } \mu\text{m}^3} = 1.44 \times 10^8$$

molecules in the cell.

At the lower concentration, a similar computation shows there are 2.87×10^6 molecules in the cell. The difference $1.44 \times 10^8 - 2.87 \times 10^6 = 1.41 \times 10^8$ is the number of molecules that have to be removed. Dividing this by 50 we see that 2.82×10^6 calsequestrin molecules are required.

5-8   **a.** Four: 1 to 2 to 4 to 8 to 16.

b. The radius of the egg is $r = 50 \text{ } \mu\text{m}$. The surface area of the egg is

$$SA = 4\pi r^2 = 4\pi (50)^2 = 3.14 \times 10^4 \text{ } \mu\text{m}^2$$

The volume of the egg is

$$V = \frac{4}{3}\pi (50)^3 = 5.24 \times 10^5 \text{ } \mu\text{m}^3$$

The volume of the morula cell is 1/16 this, or $3.275 \times 10^4 \text{ } \mu\text{m}^3$. Assuming we can model each morula cell again as a sphere, by solving the equation

$$3.275 \times 10^4 \text{ } \mu\text{m}^3 = \frac{4}{3}\pi r^3$$

we obtain the radius of the morula cell, $r = 19.85 \text{ } \mu\text{m}$. Then the surface area of the morula cell is

$$SA = 4\pi r^2 = 4\pi (19.85)^2 = 4.95 \times 10^3 \text{ } \mu\text{m}^2$$

c. It takes 3.14×10^5 molecules of dye for the egg.

d. It takes $16 \times (4.95 \times 10^3) \times 10 = 7.92 \times 10^5$ molecules for the morula cells.

e. The ratio is $50/19.85 = 2.52$. The cube of this number is 16, the ratio of the volumes.

f. The ratio is $(3.14 \times 10^4)/(4.95 \times 10^3) = 6.34$. This is the square of the ratio of the radii, 2.52.

5-9

a. $25 \times 25 \times 100 = 62{,}500 \ \mu m^3$.

b. $0.15 \times 62{,}500 = 9375 \ \mu m^3$.

c.

$$\frac{0.149 \ \text{mol}}{1 \ \text{l}} \times \frac{1 \ \text{l}}{10^3 \ cm^3} \times \frac{1 \ cm^3}{10^{12} \ \mu m^3} \times 9375 \ \mu m^3 = 1.4 \times 10^{-12} \ \text{mol}$$

d.

$$\frac{23{,}164 \ \text{cal}}{1 \ \text{mol} \cdot V} \times (1.40 \times 10^{-12} \ \text{mol}) \times (0.15 \ V) = 4.86 \times 10^{-9} \ \text{cal}$$

5-10

a. The volume of a liposome is

$$V = \frac{4}{3}\pi r^3 = \frac{4}{3}\pi (2 \ \mu m)^3 = 33.5 \ \mu m^3$$

$$[K^+] = 100 \ \text{mM} = \frac{0.1 \ \text{mol}}{1 \ \text{l}} \times \frac{1 \ \text{l}}{1000 \ cm^3} \times \frac{1 \ cm^3}{1 \times 10^{12} \ \mu m^3} = 1 \times 10^{-16} \ \text{mol}/\mu m^3$$

The number of moles of K^+ in one liposome is therefore $(1 \times 10^{-16} \ \text{mol}/\mu m^3) \times 33.5 \ \mu m^3 = 3.35 \times 10^{-15}$ mol. The number of potassium ions in one liposome is $(3.35 \times 10^{-15}) \times (6.02 \times 10^{23}) = 2.017 \times 10^9$.

b. Each time a pore opens, 3×10^6 ions leak out. So the number of pore openings required is $(2.017 \times 10^9)/(3 \times 10^6) = 672$.

The ten pores are operating simultaneously, so need to open 67.2 times. Each time requires 2.3 seconds. So in all, $67.2 \times 2.3 = 155$ seconds or a bit over two and one-half minutes is required.

5-11

a. The surface area of a mast cell is

$$SA = 4\pi r^2 = 4\pi (10.5)^2 = 1385 \ \mu m^2$$

The surface area of one vesicle is

$$SA = 4\pi (0.75)^2 = 7.07 \; \mu m^2$$

Thus 50 vesicles lead to a temporary increase in surface area of $50 \times 7.07 = 353 \; \mu m^2$. The fraction of new surface area to old surface area is $353/1385 = 0.255$, so there is about a 26% increase.

5-12

a. The volume of an egg is

$$V = \frac{4}{3}\pi r^3 = \frac{4}{3}\pi (50 \; \mu m)^3 = 5.236 \times 10^5 \; \mu m^3$$

Before fertilization,

$$[H^+] = 10^{-6.8} \, M = \frac{1.585 \times 10^{-7} \, mol}{1 \, l} \times \frac{1 \, l}{1000 \, cm^3} \times \frac{1 \, cm^3}{1 \times 10^{12} \; \mu m^3}$$

$$= 1.585 \times 10^{-22} \, mol/\mu m^3$$

The number of moles of H^+ in one egg before fertilization is therefore $(1.585 \times 10^{-22} \, mol/\mu m^3) \times (5.236 \times 10^5 \; \mu m^3) = 8.30 \times 10^{-17}$ mol. The number of hydrogen ions in one egg is $(8.30 \times 10^{-17}) \times (6.02 \times 10^{23}) = 5.0 \times 10^7$.

Note that this computation was accomplished by multiplying the concentration by 3.152×10^{14}. Thus, the number of hydrogen ions after fertilization is $10^{-7.25} \times (3.152 \times 10^{14}) = (5.623 \times 10^{-8}) \times (3.152 \times 10^{14}) = 1.77 \times 10^7$. The difference is the number of ions pumped out: $(5 \times 10^7) - (1.77 \times 10^7) = 3.23 \times 10^7$.

b. The number of calcium ions before fertilization is $10^{-8} \times (3.152 \times 10^{14}) = 3.15 \times 10^6$. The number of calcium ions after fertilization is $(5 \times 10^{-7}) \times (3.15 \times 10^{14}) = 1.58 \times 10^8$. The difference is $(1.58 \times 10^8) - (3.15 \times 10^6) = 1.55 \times 10^8$.

c. The rate is

$$\frac{3.23 \times 10^7 \, ions}{200 \, sec} = 1.62 \times 10^5 \, ions/sec$$

d. The rate is

$$\frac{1.55 \times 10^8 \text{ ions}}{200 \text{ sec}} = 7.75 \times 10^5 \text{ ions/sec}$$

e. The surface area of the cell is

$$SA = 4\pi r^2 = 4\pi(50)^2 = 3.14 \times 10^4 \ \mu\text{m}^2$$

Thus

$$\frac{1.62 \times 10^5 \text{ ions/sec}}{3.14 \times 10^4 \ \mu\text{m}^2} = 5.16 \text{ ions/}\mu\text{m}^2\text{/sec}$$

5-13   **a.** The molecular weight of the protein component of each glycophorin is (131)(130) = 17,030. This is 40% of the weight of the protein, so the entire weight is 42,575. Of this, 60% is sugar, or 25,545. Thus, one modified sugar residue has a molecular weight of about 255 daltons.

b. $(6 \times 10^5) \times 100 = 6 \times 10^7$.

c.

$$\frac{17,030 \text{ g/mol}}{6.02 \times 10^{23} \text{ molecules/mol}} = 2.83 \times 10^{-20} \text{ g/molecule}$$

This times 6×10^5 molcules yields the weight 1.7×10^{-14} grams.

CHAPTER 6

Respiration

6-1   **a.** Glucose

b. Glucose and H_2O

c. O_2

d. Glucose and H_2O

e. 100%

f. 0%

g. 44

h. 46

i. 20

6-2 The equations are completed using the numbers

Glucose: 6, 6, 6
Stearic acid: 26, 18, 18
Leucine: 15/2, 6, 5, 1

   a. For glucose, 6/6 = 1. For stearic acid, 18/26 = 0.69. For leucine, 6/(15/2) = 0.8.

6-3   **a.** Make the assumption that all of the energy in the candy actually gets to a respiratory site in the cell. The 180-calorie candy bar yields 180 kcal of energy, of which only 39%, or 70.2 kcal, is useful. They can convert

$$\frac{70.2 \text{ kcal}}{7.3 \text{ kcal/mol}} = 9.62 \text{ mol}$$

   of ATP. This is $9.62 \times (6.02 \times 10^{23}) = 5.79 \times 10^{24}$ molecules of ATP.

6-4   **a.** $4(-7.3) = -29.2$ kcal.

b. $2(-43.4) + 10(-51.7) = -603.8$ kcal.

c. $32(-7.3) + 4(-7.3) =$
$-233.6 - 29.2 = -262.8$ kcal.

d. $-233.6/-262.8 = 0.889$, so 88.9%.

   e. The glucose energy that is conserved is 262.8 kcal. The amount lost is $686 - 262.8 = 423.2$ kcal/mol. The fraction lost is $423.2/686 = 0.62$, so 62% is lost as heat.

6-5

a. At 20 kcal/kg/day, the rate is 20/24 kcal/kg/hr. For a 70 kg student, this is 58.33 kcal/hr. Dividing, we find the oxygen requirement is

$$\frac{58.33 \text{ kcal/hr}}{4.8 \text{ kcal/l}} = 12.15 \text{ l/hr}$$

b. Five times as much air is needed, or 60.8 liters.

c. Similarly for glucose,

$$\frac{58.33 \text{ kcal/hr}}{686 \text{ kcal/mol}} = 0.085 \text{ mol/hr}$$

Using the molecular weight of glucose, we find $0.085 \text{ mol/hr} \times 180 \text{ g/mol} = 15.3 \text{ g/hr}$.

d. Assume she runs for 2 hours, sleeps for 8 hours, is active for 7 hours, and is at the basal rate for 7 hours:

$$70 \left(2 \times 14.3 + 8 \times \frac{1.2}{24} + 7 \times \frac{35}{24} + 7 \times \frac{20}{24} \right) = 3150 \text{ kcal}$$

CHAPTER 7

Photosynthesis

7-1 a. There are 6.02×10^{23} molecules in a mole, so 6.02×10^{17} in a μmole. Thus, in 28 μmoles there are $28 \times (6.02 \times 10^{17}) = 1.69 \times 10^{19}$ molecules of CO_2.

b. $3 \times 1.69 \times 10^{19} = 5.07 \times 10^{19}$ ATP molecules.

c. $3 \times 5.07 \times 10^{19} = 1.52 \times 10^{20}$ H^+ ions.

7-2 a. Each mole of glucose requires 6 moles of carbon. Each mole of carbon requires 3 moles of ATP. Each mole of ATP yields 7.3 kcal of energy. Thus $6 \times 3 \times 7.3 = 131.4$ kcal/mol is required by a C_3 plant.

b. $6 \times 5 \times 7.3 = 219$ kcal/mol.

c. $6 \times 5.5 \times 7.3 = 240.9$ kcal/mol.

d. The CAM plants use the most.

7-3 a. The 3 ATP molecules require 3(7.3) kcal and the 2 NADPHs require 2(51.7) kcal, for a total of 125.3 kcal.

b. Similarly, we obtain $5(7.3) + 2(51.7) = 139.9$ kcal.

c. Photorespiration requires $2(7.3) + 2.5(51.7) = 143.85$ kcal. The total energy costs for a C_3 plant that photorespires are $125.3 + 1/3(143.85) = 173.25$ kcal.

7-4 a. The volume of a thylakoid lumen is

$$V = \pi r^2 h = \pi (0.25 \ \mu m)^2 (80 \times 10^{-4} \ \mu m) = 1.57 \times 10^{-3} \ \mu m^3$$

At pH 4, the concentration of H^+ is

$$\frac{10^{-4} \ mol}{1 \ l} = \frac{10^{-4} \ mol}{10^3 \ cm^3} \times \frac{1 \ cm^3}{10^{12} \ \mu m^3} = 10^{-19} \ mol/\mu m^3$$

So in one lumen there are $(1.57 \times 10^{-3} \ \mu m^3) \times (10^{-19} \ mol/\mu m^3) = 1.57 \times 10^{-22}$ mol. So there are $(1.57 \times 10^{-22}) \times (6.02 \times 10^{23}) = 94.5$ ions.

b. The same computation for pH 7.2 shows there are 0.095; essentially no ions.

c. The 94.5 hydrogen ions yield about 31 ATP molecules.

7-5   **a.**

$$\nu = \frac{c}{\lambda} = \frac{2.998 \times 10^8 \text{ m/sec}}{450 \times 10^{-9} \text{ m}} = 6.662 \times 10^{14} \text{ peaks/sec}$$

b. Similarly, $\nu = 4.409 \times 10^{14}$ peaks per second.

c. The energy of one photon is

$$E = \hbar\nu = (1.583 \times 10^{-34}) \times (6.662 \times 10^{14}) = 1.055 \times 10^{-19} \text{ cal}$$

so the energy in one mole is $(1.055 \times 10^{-19} \text{ cal}) \times (6.02 \times 10^{23}) = 6.35 \times 10^4 \text{ cal} =$ 63.5 kcal.

d. Similarly, $E = 42.0$ kcal.

e. Using the result of part (a),

$$\frac{6.662 \times 10^{14} \text{ peaks}}{1 \text{ sec}} \times \frac{1 \text{ sec}}{10^{12} \text{ peaks/sec}} = 666.2 \text{ peaks}$$

f. Similarly, 440.9 peaks.

7-6   **a.** The Calvin cycle requires 3 ATPs for every 2 NADPHs. To produce the 2 NADPHs requires 8 quanta. This moves 8 H^+s across the thylakoid membrane. To produce the 3 ATPs, 9 H^+s are required, so one more quantum must be used in cyclic electron transport to move it. Altogether 9 quanta yield 3 ATPs and 2 NADPHs, so 6 times this many, or 54 quanta, are required for 18 ATP + 12 NADPH. They are split so that 8/9 are used for noncyclic and 1/9 for cyclic electron transport.

b. It takes 54 quanta per molecule of glucose, so one mole of red light could produce at most 1/54 mole of glucose.

c. The 8/9 mole of quanta produces one O_2 for each 8 quanta. So (1/8)(8/9) = 1/9 mole of O_2 is evolved.

d. The Calvin cycle uses 6 molecules of CO_2 for each molecule of glucose. So 6(1/54) = 1/9 mole of CO_2 is fixed.

e. None of them would change. One quantum of visible light moves one electron.

7-7

a. One mole of photons can fix 0.1 mole of CO_2. The molecular weight of CO_2 is 44, so CO_2 weighs 44 grams per mole. Thus 4.4 grams of CO_2 are fixed.

b. Each molecule of glucose requires 6 molecules of CO_2. So $0.1/6 = 0.0167$ moles of glucose are formed.

c. The energy in 0.0167 moles of glucose is 0.0167×686 kcal/mol $= 11.46$ kcal. For red light, $11.46/41 = 0.28$, so the conversion is 28% efficient.

d. For blue light $11.46/72 = 0.16$, so the conversion is 16% efficient.

7-8

a. At 100% efficiency, $2300/12 = 191.7$ micromoles would be produced.

b. $2.4/191.7 = 0.0125$, so CAM plants are 1.25% efficient.
$20/191.7 = 0.1043$, so C_3 plants are 10.43% efficient.
$40/191.7 = 0.209$, so C_4 plants are 20.9% efficient.

7-9

a. The energy in one photon at $\lambda = 550$ is

$$E = \frac{\hbar c}{\lambda}$$

$$= \frac{(1.583 \times 10^{-34} \text{ cal sec}) \times (2.998 \times 10^8 \times 10^9 \text{ nm/sec})}{550 \text{ nm}} = 8.63 \times 10^{-20} \text{ cal}$$

Ninety percent of 2300 micromoles is 2070 micromoles. The energy they contain is $(8.63 \times 10^{-20}) \times (2.07 \times 10^3) \times (6.02 \times 10^{17}) = 107.5$ cal. This much energy is received by 1 square meter per second, so 1.075 cal is received by 100 cm^2 per second.

The number of calories required to raise 1.25 grams of water 75 degrees is $1.25 \times 75 = 93.75$. Thus it takes

$$\frac{93.75 \text{ cal}}{1.075 \text{ cal/sec}} = 87.2 \text{ sec}$$

or about 1.5 minutes.

b. To boil off the water would require $1.25 \times 540 = 675$ cal. Thus it takes

$$\frac{675 \text{ cal}}{1.075 \text{ cal/sec}} = 628 \text{ sec}$$

or about 10.5 minutes.

7-10   **a.** The energy per photon in red activating light is

$$E = \frac{\hbar c}{\lambda}$$

$$= \frac{(1.583 \times 10^{-34} \text{ cal sec}) \times (2.998 \times 10^8 \times 10^9 \text{ nm/sec})}{680 \text{ nm}} = 6.98 \times 10^{-20} \text{ cal/photon}$$

From one mole of photons, $(6.98 \times 10^{-20}) \times (6.02 \times 10^{23}) = 42.0 \times 10^3 \text{ cal} = $ 42.0 kcal of energy is obtained.

b. Similarly, the energy in one mole of light with wavelength 690 nm is 6.88×10^{-20} cal/photon or in one mole, 41.4 kcal.

c. The difference, $42.0 - 41.4 = 0.6$ kcal, is lost as heat.

d. The fraction lost as heat is $0.6/42.0 = 0.014$, so 1.4% is lost as heat.

e. The energy in one mole of blue light is computed as in (a). The result is 1.05×10^{-19} cal/photon or in one mole, 63.5 kcal. The difference, $63.5 - 41.4 = 22.1$ kcal, is lost as heat. The fraction lost as heat is $22.1/63.5 = 0.348$, so 34.8%.

CHAPTER 8

Mitosis

8-1   **a.** Attaching 15 chromosomes to microtubules takes up $220 \times 15 = 3300$ base pairs. This leaves $1.25 \times 10^7 - 3300 = 12,496,700$ base pairs, or effectively still 12.5 million. Each nucleosome and adjacent link uses $140 + 60 = 200$ base pairs. Dividing, we see that $(12.5 \times 10^6)/200 = 62,500$ nucleosomes are needed.

8-2   **a.** We need to multiply 250,000 times the number of minutes in 15 months. Assume there are 30 days in a month.

$$\frac{250,000 \text{ cells}}{\text{min}} \times \frac{60 \text{ min}}{1 \text{ hr}} \times \frac{24 \text{ hr}}{1 \text{ day}} \times \frac{30 \text{ days}}{1 \text{ month}} \times 15 \text{ months} = 1.62 \times 10^{11}$$

nerve cells.

8-3   **a.** One DNA polymerase can catalyze at the rate of 100 bases per second. The 6 billion base pairs, or 12 billion bases, must be catalyzed in 9 hours or 32,400 seconds. They are therefore catalyzed at the rate of $(12 \times 10^9)/32,400 = 370,370$ base pairs per second. Dividing by 100 we see that 3700 copies of DNA polymerase are needed.

8-4   **a.** It takes $1200/2 = 600$ times as long.

b. The chromosomes can only begin to duplicate after fusion has taken place. So $5500 - 1200 = 4300$ seconds or 1.2 hours is the maximum time available.

8-5   **a.** $2 \times (12 \times 10^9)/36 = 6.67 \times 10^8$ molecules of glucose.

b. Assuming the chromosomes have equal lengths, $(6.67 \times 10^8)/46 = 14.5 \times 10^6$ or 14.5 million.

8-6   **a.** The surface area $SA = 4\pi r^2 = 4\pi (3.5)^2 = 154 \ \mu m^2$.

b. So there are $154 \times 11 = 1700$ pores.

c. If 1700 pores transport 10^6 histones in 3 minutes, then one pore transports $10^6/1700$ histones in 3 minutes or $10^6/(1700 \times 3) = 200$ histones in 1 minute.

8-7

a. The number of hours in a week is $7 \times 24 = 168$. A plant or animal cell dividing every 18 hours thus divides $168/18 = 9.33$ times. One cell grows to $2^{9.33} = 644$ cells. One cell that doubles every 14 hours divides 12 times to create $2^{12} = 4096$ cells. A yeast cell that divides every 2 hours divides 84 times, into $2^{84} = 1.93 \times 10^{25}$ cells. For bacteria cells, we must find the number of minutes in a week:

$$1 \text{ wk} \times \frac{7 \text{ days}}{1 \text{ wk}} \times \frac{24 \text{ hr}}{1 \text{ day}} \times \frac{60 \text{ min}}{1 \text{ hr}} = 10{,}080 \text{ min}$$

Dividing by 20, we find that 504 divisions have taken place. So 1 bacterium has grown to $2^{504} = 5.24 \times 10^{151}$.

8-8

a. The number of divisions x to form 64 cells is the solution of the equation $2^x = 64$, so $x = 6$.

b. After 12 divisions there are $2^{12} = 4096$ cells.

c. In each case, we want the number of cells formed during the time interval divided by the length of the time interval.
$(2 - 1)/(1/2 - 0) = 2$ cells per hour.
$(64 - 2)/(4 - 1/2) = 17.71$ cells per hour.
$(10{,}000 - 64)/(6 - 4) = 4968$ cells per hour.
$(30{,}000 - 10{,}000)/(10 - 6) = 5000$ cells per hour.
$(80{,}000 - 30{,}000)/(19 - 10) = 5556$ cells per hour.
$(170{,}000 - 80{,}000)/(32 - 19) = 6923$ cells per hour.
$(10^6 - 170{,}000)/(110 - 32) = 10{,}641$ cells per hour.

d. Between somite formation and the feeding tadpole. The rate of cell production keeps increasing during the first 110 hours.

8-9

a. The actual diameter is $4/1000 = 0.004$ mm or 4 μm.

b. Using $V = \frac{4}{3}\pi r^3$,

$$V = \frac{4}{3} \times \pi \times 2^3 \mu\text{m}^3 = 33.5 \mu\text{m}^3$$

c. One-sixteenth of this is the volume of one morula cell, or $2.09 \ \mu\text{m}^3$.

d. Assuming the morula cells are also spheres, their radius r is the solution to the equation

$$V = 2.09 = \frac{4}{3}\pi r^3$$

so $r = 0.793 \mu\text{m}$ and the diameter is $1.59 \ \mu\text{m}$.

8-10   **a.** Assume the microtubules have length $2 \ \mu\text{m}$ and that 25 of them are attached to each chromosome. Then the total length of all the microtubules is $46 \times 25 \times 2 \ \mu\text{m} = 2300 \ \mu\text{m}$. To find the length of the protofilaments in nanometers, $2300 \ \mu\text{m} \times 13 \times 1000 \ \text{nm}/\mu\text{m} = 29{,}900{,}000 \ \text{nm}$. Thus $29{,}900{,}000/8 = 3{,}740{,}000$ tubulin subunits are released.

b. Assume the process of releasing 3,740,000 tubulin subunits takes 6 minutes. Then $3{,}740{,}000/6 = 623{,}000$ are released per minute, or about 10,000 per second.

8-11   **a.** An individual chromosome has $(3 \times 10^9)/23 = 1.3 \times 10^8$ base pairs. Half of them, or 6.5×10^7, are AT pairs with two H bonds, and the other half are GC pairs with three H bonds. So in all there are $(6.5 \times 10^7 \times 2) + (6.5 \times 10^7 \times 3) = 3.25 \times 10^8$ H bonds.

b. $(3.25 \times 10^8)/110 = 2.95 \times 10^6$.

8-12   **a.** In the chlorplasts, there are $171 \times 32 \times 150 = 8.2 \times 10^5$ kbp. In the cell there are 3.6 million kbp. The fraction is $(8.2 \times 10^5)/(3.6 \times 10^6) = 0.228$, so 23% are in the chloroplasts.

b. In the mitochondria, there are $10 \times 10 \times 1000 = 10^5$ kbp. The fraction is $10^5/(3.6 \times 10^6) = 0.028$, so 2.8% are in the mitochondria.

c. This leaves $100 - 23 - 2.8 = 74\%$ in the nucleus.

CHAPTER 9

Mendelian Genetics

9-1   **a.**

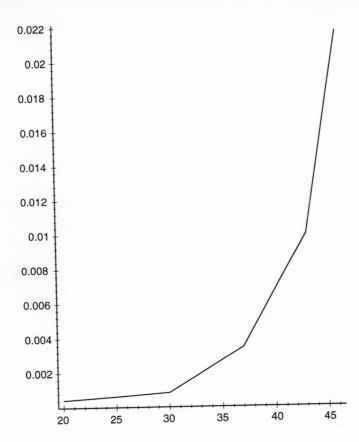

b. Exponentially.

c. $(1/46) \div (1/2300) = 50$ times.

9-2   **a.** For each of two choices for the first gene, there are two for the second, and so on, for a total of $2^7 = 128$.

b. $128 \times 128 = 16,384$.

c. For each gene there are three possibilities: AA, Aa, or aa. There are three choices for the first gene, three for the second, and so on, for a total of $3^7 = 2187$ different genotypes.

d. For each gene there are two possibilitites: dominant or recessive. Hence, there are 128 different phenotypes.

9-3
a. All are Aa; dominant.

b. All are AA; dominant.

c. All are aa; recessive.

d. 1:1, dominant Aa to recessive aa.

e. All are AA or Aa; dominant.

f. 3:1, dominant AA or Aa to recessive aa.

g. 2:1, dominant Aa to recessive aa.

h. All are AA or Aa; dominant.

i. 1:2:1, AA to Aa to aa.

9-4
a. We label the alleles of each gene $+$ and $-$. Each person then has one of the three combinations $++$, $+-$, or $--$ at each of three sites. So there are $3^3 = 27$ possible genotypes.

b. The phenotype depends only on the number of occurrences of $+$. This can range from 0 to 6. So there are seven possible phenotypes.

9-5
a. 4: CC, CC^{ch}, CC^h, Cc

b. 3: $C^{ch}C^{ch}, C^{ch}C^h, C^{ch}c$

c. 2: C^hC^h, C^hc

d. 1: cc

e. One parent must have had a C^{ch} gene for it to appear in an offspring. It couldn't have been the one with the C^h or that rabbit would have appeared chinchilla. Hence, one parent is CC^{ch} and the other is either C^hC^h or C^hc. The cross then is

	C	C^{ch}
C^h	CC^h	$C^{ch}C^h$
C^h or c	CC^h or Cc	$C^{ch}C^h$ or $C^{ch}c$

The ratio expected is 1:1, full color to chinchilla.

9-6
a. Of the 16 possibilities of the cross, 9 are $Na\_\_ Le\_\_$ and are tall, 3 are $Na\_\_ lele$ and are short, and 4 are $nana \_\_\_\_$ and are extremely short. The ratio is 9:3:4.

9-7
a. Of the 16 possibilities of the cross, 1 is $CCPP$, 2 are $CCPp$, 1 is $CCpp$, 2 are $CcPP$, 4 are $CcPp$, 2 are $Ccpp$, 1 is $ccPP$, 2 are $ccPp$, and 1 is $ccpp$.

b. Nine are $C\_\_ P\_\_$, purple; 3 are $C\_\_ pp$, bronze; 4 are $cc\_\_\_\_$, brown.

9-8   **a.** No, if you have one defective chromosome you have the syndrome.

b. Letting A stand for a normal chromosome and a for a defective one, AA is normal and has probability 1/4.

c. Aa has probability $2/4 = 1/2$.

d. aa has probability 1/4.

9-9   **a.** Create the Punnett square

	ear		
		$2P$	$2p$
tassel	P	PPP	Ppp
	p	PPp	ppp

The phenotypes purple and white occur in the ratio 3:1.

b.

	ear		
		$2P$	$2p$
tassel	p	PPp	ppp

Purple and white occur in the ratio 1:1.

c.

	ear	
		$2p$
tassel	P	Ppp
	p	ppp

Purple and white occur in the ratio 1:1.

d. Using the square from (a), the colors would appear in the ratio 1:1:1:1. In (b), they would occur in the ratio 1:1, violet to white. In (c) they would appear in the ratio 1:1, red to white.

9-10

a. Crossing *GGRR* with *ggrr* must yield *GgRr* for the offspring.

b. Crossing *GgRr* with *ggrr* would yield *GgRr*, *Ggrr*, *ggRr*, and *ggrr*, in equal numbers. Thus, there would be 25% each of grey–regular wing, grey–vestigial wing, black–regular wing, and black–vestigial wing flies.

c. Autosomal linkage, because it is unrelated to sex differences.

d. The gray–vestigial and black–regular offspring are recombinant, since they do not display the type of either parent. They constitute 20% of the offspring, so the genes are 20 map units apart.

e. Because the genes for body color and wing length are 20 units apart and eye color is 10 units from each, it must be in the middle. Thus the order is either body, eye, wing or wing, eye, body.

9-11

a. There are two choices for each of genes A, B, C, and D. Hence $2^4 = 16$ different gametes can be produced.

b. There are three choices for each gene–pure dominant, hybrid, and recessive. Hence $3^4 = 81$ different genotypes can occur.

c. There are two different choices for each gene–dominant or recessive. Hence $2^4 = 16$ different phenotypes can appear.

d. Because genes B and D appear to assort independently, we assume they are the farthest away. The order is *BACD*.

e. $31 - 17 = 14$ centimorgans.

9-12

a. The distance from A to G in women is 129 map units and in men is 78 map units. Thus the woman's is $129/78 = 1.65$ times longer than a man's, or is 65% longer. Assuming the likelihood of a crossover is directly proportional to the length, a crossover in a woman is 65% more likely than in a man.

b. The difference in length is 51 map units or 51,000 kilo base pairs.

CHAPTER 10

Population Genetics

10-1

a. For flower color,

$$\chi^2 = \frac{(705 - 696.75)^2}{696.75} + \frac{(224 - 232.25)^2}{232.25} = 0.3907$$

This is less than 3.841, so the hypothesis is not rejected.

b. Neither are any of the others. The values of χ^2 are 0.3497 for flower position, 0.0150 for seed color, 0.2629 for seed shape, 0.0635 for pod shape, 0.4505 for pod color, and 0.6065 for stem length.

> Did Mendel fudge his data? R. A. Fisher analyzed the results of Mendel's experiments in hybridization. Fisher noted that the data were unusually close to theoretically expected outcomes. He says, "The data have evidently been sophisticated systematically, and after examining various possibilities, I have no doubt that Mendel was deceived by a gardening assistant, who knew only too well what his principal expected from each trial made." Fisher used chi-square tests and concluded that only about a 0.00004 probability exists for such close agreement between expected and reported observations.
> —Mario F. Triola, *Elementary Statistics*, Addison-Wesley

10-2

a. All four entries are 250.

b.

$$\chi^2 = \frac{69^2}{250} + \frac{76^2}{250} + \frac{59^2}{250} + \frac{66^2}{250} = 73.5$$

c. Because $73.5 > 7.815$ we reject the hypothesis.

d. $\chi^2 = 5.37 < 7.815$, so this time we do not reject the hypothesis.

e. If we crossed with AABB, all the offspring would display the dominant phenotype.

10-3

a. 381.

b. 9:3:3:1.

c. 214.3, 71.4, 71.4, 23.8.

d. $\chi^2 = 134.7$.

e. Because $134.7 > 7.815$ we reject the hypothesis.

f. 42/381 = 0.11, so 11%.

g. 148.59, 41.91, 41.91, 148.59.

h. $\chi^2 = 204.2$. We reject this hypothesis also.

So while it seems clear the genes are linked, likely they are not equally viable or perhaps the sample size is just too small to determine crossover rates.

10-4   **a.** The frequency of B is $p = 0.7$ and the frequency of b is 0.3. So in a cross, the frequency of BB is $p^2 = 0.49$, of Bb is $2pq = 0.42$, and of bb is $q^2 = 0.09$. Those displaying trait B are $0.49 + 0.42$ or 91% of the population.

b. Carriers, who do not display trait b, are the hybrids, thus 42%.

c. These are the individuals with genes bb; thus 9%.

d. With 16% of the people being left-handed, $q^2 = 0.16$ and $q = 0.4$. Therefore $p = 1 - 0.4 = 0.6$. The homozygous right-handers have frequency $p^2 = 0.36$. These constitute the fraction

$$\frac{0.36}{0.84} = 0.4286$$

or 43% of the right-handed people.

e. The rest, 57%, are heterozygous right-handers.

f. If 1% display the trait, $q^2 = 0.01$ and $q = 0.1$. So $p = 1 - q = 0.9$ and $2pq = 0.18$. Thus, even though only 1% display the trait, 18% are carriers. This is the probability that any randomly chosen person is a carrier. Your being a carrier has no influence on whether or not your classmate is a carrier unless you are related.

g. These are the remaining 82%.

h. Two out of 200,000 make up a freqency of 1×10^{-5}. So $q = \sqrt{1 \times 10^{-5}} = 3.16 \times 10^{-3}$. Subtracting, we find $p = 0.9968$. The heterozygous flowers are not true-breeding, and their frequency is $2pq = 6.3 \times 10^{-3}$. So 0.63% are not true-breeding.

i. If 68% display the dominant trait, then 32% do not. So $q^2 = 0.32$, $q = 0.566$, and the frequency of the dominant allele is $p = 1 - 0.566 = 0.434$. Note that even though a majority of the population displays the dominant trait, its frequency is less than half.

10-5   **a.** Assume the alleles P and p are equally likely to be handed down to the next generation. Then the distribution of PP, Pp, and pp is 1/4 to 1/2 to 1/4. Note that we have identified Pp and pP as being the same.

b. In the set {*PP, Pp, pP, pp*} there are exactly four *P*s and exactly four *p*s. In the next generation they are still equally likely to be handed down, so the distribution is the same.

c. Starting with homozygous purple flowers, only *PP* can ever be handed down. So the distribution is 1 to 0 to 0.

d. This is the same as part (c).

e. In the initial population, the frequency of *P* is $0.6 + 0.2 = 0.8$ and the frequency of *p* is 0.2. In the first generation, the ratio of p^2 to $2pq$ to q^2 is $0.64:0.32:0.04$.

f. The second and subsequent generations are distributed the same as the first.

10-6   **a.** With $p = 0.957$ and $q = 0.043$, $2pq = 0.082$, so 8.2% are heterozygous.

b. $q^2 = 0.00185$, so 0.185% demonstrate type *B*.

c. There are 15 genotypes: *AA, AB, AC, AD, AE, BB, BC, BD, BE, CC, CD, CE, DD, DE, EE*.

CHAPTER 11

Molecular Genetics

11-1   **a.** The fraction that is changed each day is $(1 \times 10^4)/(6 \times 10^9) = 1.67 \times 10^{-6}$. So the percentage changed per day is 0.000167%.

b. At 10,000 per day, in 365 days, 3.65×10^6 changes occur. If 1 survived, the fraction that survive would be $1/(3.65 \times 10^6) = 2.74 \times 10^{-7}$ and the percentage, 0.0000274%. If 5 survived, the fraction that survive would be 5 times as great, or 0.000137%.

11-2   **a.** We make the assumption that there is a linear relationship between the quantity of toxin and the mutations. Then $7057 \div 0.006 = 1,176,167$ nmoles or 1.18 mmoles.

11-3   **a.** Assume that the cpDNA rate equals 1/3 the mtDNA rate in plants. Then the cpDNA rate equals 1/6 the nucDNA rate in plants. The nucDNA rate in plants equals that in animals equals 1/5 the animal mtDNA rate. So the plant mtDNA rate equals 1/30 the animal mtDNA rate. So the plant mtDNA mutation rate is 30 times slower.

Without the assumption we made of equality, we would have to say the plant mtDNA mutation rate is *at least* 30 times slower.

11-4   **a.** We must solve $1 - q^n = 0.9$, or equivalently, $q^n = 0.1$, for n, first with $q = 0.99951$ and again for $q = 0.999999$. For the first, using logarithms base 10,

$$n = \frac{\log 0.1}{\log 0.99951} = 4700 \text{ kernels}$$

Similarly for the lower frequency (one mutant in a million), to be 90% certain of finding one we would need to inspect 2,300,000.

11-5   **a.** We must solve $1 - q^n = 0.9$, or equivalently, $q^n = 0.1$, for n, for three values of q. Under natural conditions, $p = 4 \times 10^{-7}$, $q = 0.9999996$. Using logarithms base 10,

$$n = \frac{\log 0.1}{\log 0.9999996} = 5,760,000 \text{ spores}$$

Similarly for the second frequency, 92,100 spores. For the third, 6140 spores.

11-6   **a.** Sixteen percent of 1 million spores, or 160,000 survive. The fraction that mutate is $259/160{,}000 = 0.0016$ or 0.16%.

b. Using $p = 0.0016$, $q = 0.9984$. We must solve $1 - q^n = 0.9$, or equivalently, $q^n = 0.1$, for n. Using logarithms base 10,

$$n = \frac{\log 0.1}{\log 0.9984} = 1440 \text{ spores}$$

c.

$$n = \frac{\log 0.05}{\log 0.9984} = 1870 \text{ spores}$$

CHAPTER 12

Protein Synthesis and Lifetime

12-1   **a.** 146 amino acid codons × 3 nucleotides/amino acid codon = 438 nucleotides. 438 nucleotides × 2 ATPs/nucleotide = 876 ATPs (or equivalent TTP, GTP, or CTP).

b. 146 amino acid attachments × 2 ATPs/attachment = 292 ATPs.

c. 876 + 292 = 1168 ATPs.

d. 1 for initiation +(2)(145) for codon migration + 1 for termination = 292 GTPs.

12-2   **a.** There are 360,000 × 10 × 0.75 = 2.7×10^6 active ribosomes, each engaged in synthesizing a protein.

b. If we assume 7.5 ribosomes are reading 5 mRNAs, 37.5 proteins are in the process of creation.

c. With 7.5 ribosomes reading 12,000 mRNAs, 90,000 proteins are in the process of creation.

d. The ratio is 90,000:37.5 or 2400:1.

12-3   **a.** Solving for K with $t_{1/2} = 24$, $K = 0.0289$ per hour or 4.82×10^{-4} per minute.

b. Similarly $K = 0.347$ per minute.

c. and d. After 5 half-lives, the fraction remaining is $1/32 = 0.03125$ or 3.125%.

e. The ratio is $0.346/(4.82 \times 10^{-4}) = 720$. So ARG-amino-end protein degradation is 72,000% faster than MET-amino-end degradation.

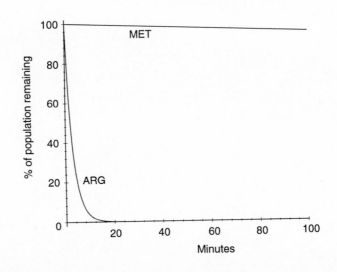

h. One hundred twenty days is 120 half-lives for a valine-terminated protein. The initial amount would be reduced to $1/2^{120} = 7.5 \times 10^{-37}$ of the original amount. This level is essentially zero, but is almost detectable. For a leucine-terminated protein, this level would be reached in $120 \times 2 = 240$ minutes. After 4 hours the level would be undetectable, so it would not last 120 days. It is in fact valine.

12-4

    a. One ribosome takes $145 \times 0.06 = 8.7$ seconds to build one hemoglobin polypeptide chain. Thus in 3 minutes, or 180 seconds, $180/8.7 = 20.69$ hemoglobin polypeptide chains can be made by the ribosome. The mRNA has 145 codons or 435 nucleotides. With one ribosome every 80 nucleotides of the 3 copies of mRNA, there are $(3 \times 435)/80 = 16.3$ ribosomes working. Thus, they can make $16.3 \times 20.69 = 337$ hemoglobin polypeptide chains.

CHAPTER 13

Biotechnology

13-1 For bacterial plasmids,

$$\frac{6 \times 10^9 \text{ bp}}{15,000 \text{ bp/plasmid}} = 400,000 \text{ plasmids}$$

For phages, from 240,000 to 1,200,000.
For cosmids, from 133,000 to 170,000.
For YACs, from 20,000 to 30,000.

13-2   **a.** Two H bonds are required for AT and three for CG. The first segment has 6 AT pairs, for a total of 12 H bonds, and 6 CG pairs for a total of 18 H bonds. In total there are 30 H bonds.

b. A similar calculation shows the second segment has 33 H bonds.

c. The first has fewer bonds to break, so will become single-stranded more easily.

13-3   **a.** Gly is coded in 4 ways. Val is also coded in 4 ways. Thus the pair can be coded in any of $4 \times 4 = 16$ ways, and as the sequence becomes longer, the same principle holds. Consequently, segment A can be coded in any of $4 \times 4 \times 6 \times 4 \times 6 \times 4 = 9216$ ways.

b. For segment B, $1 \times 2 \times 1 \times 2 \times 2 \times 1 = 8$ ways.

c. Use segment B. Then only 8 searches are needed, rather than 9216.

CHAPTER 14

Evolution

14-1

a. The difference is greater between the original and the high-protein progeny.

b. The ratio of increase is $(19.4 - 10.9)/10.9 = 0.78$, so there was a 78% increase in the protein content of the high-protein progeny. The ratio of decrease is $(10.9 - 4.9)/10.9 = 0.55$, so there was a 55% decrease in the low.

c. The average rate of increase over 50 years in the high-protein seeds is $(19.4 - 10.9)/50 = 0.17$, or 17% more protein per year. The rate of increase in the low-protein seeds is $(4.9 - 10.9)/50 = -0.12$, so there was a 12% decrease in the amount of protein per year.

14-2

a. At 3 generations back you have $2^3 = 8$ ancestors. The most likely fraction of your DNA from any one of them is 1/8.

b. The theoretical maximum is 1/2, if the genes your father passed on were the half he inherited from his mother, and the genes she passed on were the half she inherited from her father.

c. The answer depends on the year in which you work this problem, but in 1999, for example, there had been $(1999 - 1776)/25 = 9$ generations, so you have had $2^9 = 512$ ancestors in that time. Since 1066 you have had 2^{37}.

14-3

a. 150,000 seconds $= 150,000/60/60/24 = 1.736$ days, so about 1.7 days.

b. There are $60 \times 60 \times 24 \times 365 = 3.154 \times 10^7$ seconds in a year. So 65 million seconds divided by $3.154 \times 10^7 = 2.1$ years.

c. 2.5 years.   d. 13 years.   e. 111 years.

14-4

a. For the monkey, $26,000,000/8 = 3,250,000$ years; the mouse $80,000,000/25 = 3,200,000$; the chicken $275,000,000/45 = 6,110,000$; the frog $330,000,000/70 = 4,710,000$; the lamprey $450,000,000/125 = 3,600,000$.

b. For the monkey, $8/26 = 0.31$, the mouse $25/80 = 0.31$, the chicken $45/275 = 0.16$, the frog $70/330 = 0.21$, the lamprey $125/450 = 0.28$.

c. The average of the rates in (b) is 0.254. The closest is the lamprey.

d. For the monkey, $8/146 = 0.055$, or 5.5%. For the lamprey, $125/146 = 0.86$ or 86%.

14-5   **a.** The data used below are in each case the averages of the values given in the table: $(-2.7, 600), (-2.0, 650), (-1.2, 950), (-0.09, 1400), (-0.035, 1400), (-0.05, 1400)$.

b. $\sum x_i = -6.075, \sum y_i = 6400, \sum x_i^2 = 12.74, \sum x_i y_i = -4305, m = 330, b = 1401$.

c. Using the slope of the regression line,

$$\frac{\text{change in brain size}}{\text{elapsed time}} = 330 \text{ cm}^3/\text{million years}$$

So the elapsed time is $100/330 = 0.3$ million years.

d. From *Australopithecus* to *Homo habilis*, $(650 - 600)/600 = 0.0833$, so 8.3%. From *Homo habilis* to *Homo erectus*, $(950 - 650)/650 = 0.462$, so 46%. From *Homo erectus* to *Homo sapiens*, $(1400 - 950)/950 = 0.474$, so 47%.

14-6   **a.** For human and gorilla, $5/75 = 0.066$, so 6.7%. For human and orangutan, $8/75 = 0.1066$, so 11%. For gorilla and orangutan, $9/75 = 0.12$, so 12%.

b. For human and gorilla, $13/225 = 0.057$, so 5.7%. For human and orangutan, $19/225 = 0.084$, so 8.4%. For gorilla and orangutan, $26/225 = 0.1155$, so 12%.

c. Gorillas.

d. Humans.

e. Humans and gorillas have five differences in amino acids in 10 million years, so $10/5 = 2$ million years per difference. For humans and orangutans, $10.5/8 = 1.31$ million years and for gorillas and orangutans, 1.17 million years. If we average these we obtain $(2 + 1.31 + 1.17)/3 = 1.5$ million years as an approximate time for an amino acid mutation. Similarly, for nucleotides we obtain 0.57 million years.

14-7   **a.** If it takes 500 million years to effect a change in a 100 amino acid sequence, it should take proportionally less time for a change in a longer sequence. Thus, it takes $(100/102) \times 500 = 490$ million years to make a change in a sequence of length 102. So two changes tell us there were 2×490 million years, or almost 1 billion years.

b. Using 108 as the average length of cytochrome c, the time required for one change is $(100/108) \times 21 = 19.4$ million years. So the time to a common ancestor of humans and rhesus monkeys is about 20 million years, dogs $13 \times 19.4 = 250$ million years, rattlesnakes 390 million years, tuna 600 million years, and mung beans 830 million years.

c. The ratio $(500 - 21)/21 = 22.8$, so cytochrome c mutates 2280% faster.

14-8   **a.** The average of the six rates is 2.0 changes per billion years.

b. At 0.01 change per billion years it would take 100 billion years. This is much longer than the age of the earth, so histone 4 does not make a good evolutionary clock.

c. At 0.45 change per billion years, for 1 change it takes $1/0.45 = 2.22$ billion years. In 2.22 billion years, fibronectin would make $9 \times 2.22 = 20$ changes.

d. In a billion years, the number of changes in these proteins sums to 11.92. Thus in 0.002 billion years, we would expect $0.002 \times 11.92 = 0.02$ change.

14-9   **a.** Exponential decay is governed by the equation

$$A = A_0 e^{-rt}$$

where t is time, r is the decay rate, A_0 was the original amount present at time $t = 0$, and A is the amount present at time t.

Using the fact that half the original amount of U^{238} is present when $t = t_{1/2} = 4.5 \times 10^9$, we can solve for the decay rate r:

$$\frac{1}{2} A_0 = A_0 e^{(4.5 \times 10^9)r}$$

to find $r = -\ln 2/(4.5 \times 10^9) = -1.54 \times 10^{-10}$.

At the time $t = 10^9$ when the first bacteria became fossilized,

$$A = A_0 e^{(-1.54 \times 10^{-10})(10^9)} = A_0 \times 0.857$$

So 86% of the U^{238} remained.

b. Vascular plants appeared at time $t = 4.5 \times 10^9 - 4 \times 10^8 = 4.1 \times 10^9$.

$$A = A_0 e^{(-1.54 \times 10^{-10})(4.1 \times 10^9)} = A_0 \times 0.532$$

So 53% of the U^{238} remained.

c. The time when modern man appeared is so recent on this scale that to the number of significant figures we are allowed to keep we cannot distinguish this short a period. Almost 50% still remained.

14-10   **a.** One generation has an average duration of $(4 \times 365)/10,000 = 0.146$ day or $0.146 \times 24 = 3.5$ hr.

b. $(0.58 \times 10^{-15}) - (0.35 \times 10^{-15}) = 2.3 \times 10^{-16}$ l.

c. $(0.48 \times 10^{-15}) - (0.35 \times 10^{-15}) = 1.3 \times 10^{-16}$ l.

d. $(0.23 \times 10^{-15})/10,000 = 2.3 \times 10^{-20}$ l/generation.

e. $(0.23 \times 10^{-15})/4 = 5.8 \times 10^{-17}$ l/yr.

f. $(0.13 \times 10^{-15})/300 = 4.33 \times 10^{-19}$ l/generation.

g. Three hundred generations lasts $300 \times 0.146 = 43.8$ days or $43.8/365 = 0.12$ yr, so the rate of increase is $(0.13 \times 10^{-15})/0.12 = 1.08 \times 10^{-15}$ l/yr.

14-11   **a.** There were $(50 \times 10^6)/25 = 2$ million generations.

b. The sunflower speciated $100/5 = 20$ times as fast as goat's beard, and $(50 \times 10^6)/5 = 10$ million times as fast as the sycamores.

c. The sunflower speciated in $100/5 = 20$ times fewer generations than goat's beard, and in $(2 \times 10^6)/5 = 400,000$ times fewer generations than sycamores.

14-12   **a.** In 200 million years at 2.3 cm/yr, $(2.3 \text{ cm/yr}) \times (200 \times 10^6 \text{ yr}) = 4600$ km. Thus, the motion must have been slightly faster.

b. The average speed is

$$\frac{5583 \text{ km}}{200 \times 10^6 \text{ yr}} \times \frac{10^5 \text{ cm}}{1 \text{ km}} = 2.8 \text{ cm/yr}$$

c. Using 2.8 cm/yr, in 135 million years the distance was $2.8 \times (135 \times 10^6) = 3800$ km.

14-13   **a.** The depth would be $(90 \times 10^6) \times 0.7 = 63 \times 10^6$ mm, 63 km, or $63 \times 0.62 = 39$ miles.

b. The gingko had $(150 \times 10^6) - (0.2 \times 10^6) = 149.8 \times 10^6$ years before humans, or $(149.8 \times 10^6)/20 = 7,490,000$ generations before humans, and $200,000/20 = 10,000$ generations after. Similarly the dawn redwood had 4,490,000 before, the wollenia pine had 9,990,000 before, and they each had 10,000 generations after.

CHAPTER 15

Structure

15-1   **a.** The initial diameter is 1/435 cm and the initial radius is half that: $r = 0.00115$. Using the formula for the volume of a sphere, $V = \frac{4}{3}\pi r^3$, the initial volume is

$$V = \frac{4}{3}\pi \times (0.00115)^3 = 6.37 \times 10^{-9}\ \text{cm}^3$$

Assuming the eight cells are the same size, each has volume $V = 7.96 \times 10^{-10}\ \text{cm}^3$. The radius of the larger sphere is $r = 2.5/2/435 = 0.00287$ cm. The volume of the larger sphere is

$$V = \frac{4}{3}\pi \times (0.00287)^3 = 9.90 \times 10^{-8}\ \text{cm}^3$$

Dividing this by the volume per cell we obtain $(9.90 \times 10^{-8})/(7.96 \times 10^{-10}) = 124$ cells.

15-2   **a.** For each division of quiescent cells there are $430/13 = 33$ divisons of root cap cell precursors.

b. One cell gives rise to $2^{33} = 8.6 \times 10^9$ or about 8.6 billion cells.

15-3   **a.** If the diameter grows at 75 μm/hr, the radius grows at 37.5 μm/hr. The distance from the center to the edge is 12.5 μm. To travel 12.5 μm at 37.5 μm/hr takes $12.5/37.5 = 1/3$ hr or 20 minutes.

b. The distance from the center to the far end is $8700/2 = 4350$ μm. To travel 4350 μm at 37.5 μm/hr takes 116 hours or 4.83 days.

c. It takes $116/(1/3) = 348$ times longer.

15-4   **a.** In the central zone, $288/40 = 7.2$ times as long. In the peripheral zone, $157/30 = 5.23$ times as long.

b. For rapid dividers, $40/30 = 1.33$ times as fast. For slow dividers, $288/157 = 1.83$ times as fast.

c. For a rapid divider in the central zone, $40 \times 10 = 400$ hours or 16.7 days. For a rapid divider in the peripheral zone, $30 \times 10 = 300$ hours or 12.5 days. For a slow divider in the central zone, $288 \times 10 = 2880$ hours or 120 days. For a slow divider in the peripheral zone, $157 \times 10 = 1570$ hours or 65.4 days.

15-5   a. Assume the vessel length is 75 μm. First convert 75 μm to feet:

$$75\mu m \times \frac{10^{-6} \text{ m}}{1\mu m} \times \frac{3.28 \text{ ft}}{1 \text{ m}} = 2.46 \times 10^{-4} \text{ ft}$$

So in 50 ft we have $50/(2.46 \times 10^{-4}) = 203{,}000$ cells.

b. Similarly we convert 1 mm to feet:

$$1 \text{ mm} \times \frac{10^{-3} \text{ m}}{1 \text{ mm}} \times \frac{3.28 \text{ ft}}{1 \text{ m}} = 3.28 \times 10^{-3} \text{ ft}$$

In 50 feet we have $50/(3.28 \times 10^{-3}) = 15{,}200$ cells.

15-6   a. Assuming the average, the minor axis has length $(3 + 12)/2 = 7.5$ μm and the major axis has length $(10 + 40)/2 = 25$ μm. We compute the area of a stomate:

$$A = \pi ab = \pi \times \frac{7.5}{2} \times \frac{25}{2} = 147\mu m^2$$

Assuming an average of $(1000 + 60{,}000)/2 = 30{,}500$ stomates per cm^2, we compute they occupy $3.05 \times 10^4 \times 147 = 4.48 \times 10^6 \mu m^2$ of the 1 $cm^2 = 10^8 \mu m^2$ area. The ratio is $(4.48 \times 10^6)/10^8 = 0.0448$. So about 4.5% of the area is occupied by stomates.

15-7   a. Assume that the rate of water vapor loss is the same on the upper epidermis as on the lower epidermis. There are an average of 8000 stomates/cm^2. They lose water at the rate of 700 g/m^2/hr or 0.7 g/cm^2/hr. So one stomate loses $0.7/8000 = 8.75 \times 10^{-5}$ g/hr or $(8.75 \times 10^{-5})/60 = 1.46 \times 10^{-6}$ g/min.
Now we convert from grams to moles to molecules:

$$\frac{1.46 \times 10^{-6} \text{ g}}{18 \text{ g/mol}} = 8.11 \times 10^{-8} \text{ mol}$$

or $(8.11 \times 10^{-8} \text{ mol}) \times (6.02 \times 10^{23} \text{ molecules/mol}) = 4.9 \times 10^{16}$ molecules per minute.

15-8

a. If the radius of the outer circle is R and that of the inner circle is r, the area of the ring is $\pi R^2 - \pi r^2 = \pi(R^2 - r^2)$.

b. The original circumference was 15 cm. The circumference C is related to the radius r by $C = 2\pi r$. So $r = 2.387$ cm, and without the bark the initial radius of the tree (at year 0, when the ten years start) is $r_0 = 2.387 - 0.5 = 1.89$ cm. Each year it increases by 0.6 cm. The areas in the table below are calculated using the formula in (a).

c. The relative increase in the area of the ring during the second year is $(10.52 - 8.26)/8.26 = 0.274$, so the percentage increase is 27.4%. The rest of the computations are given in the table.

d. The graph is curved.

Year	Radius	Area	Percentage Increase
0	1.89		
1	2.49	8.26 cm^2	
2	3.09	10.52 cm^2	27.4
3	3.69	12.78 cm^2	21.5
4	4.29	15.04 cm^2	17.7
5	4.89	17.30 cm^2	15.02
6	5.49	19.57 cm^2	13.12
7	6.09	21.83 cm^2	11.55
8	6.69	24.09 cm^2	10.35
9	7.29	26.35 cm^2	9.38
10	7.89	28.61 cm^2	8.58

15-9

a. The circumference C is related to the radius r by $C = 2\pi r$. So if $C = 2.5$ ft, $r = 0.3979$ ft. Converting this to micrometers,

$$r = 0.3979 \text{ ft} \times \frac{30.5 \text{ cm}}{1 \text{ ft}} \times \frac{10^4 \mu m}{1 \text{ cm}} = 1.21 \times 10^5 \mu m$$

Thus the area of the cross section is $A = \pi r^2 = \pi \times (1.21 \times 10^5)^2 = 4.60 \times 10^{10}$ μm^2. The area of one cell is $50^2 = 2500$ μm^2. So there are $(4.60 \times 10^{10})/2500$, or over 18 million cells.

b. The thickness of the slab is

$$4 \text{ in} \times \frac{2.54 \text{ cm}}{1 \text{ in}} \times \frac{10^4 \mu m}{1 \text{ cm}} = 1.016 \times 10^5 \mu m$$

So the volume of the slab is given by

$$V = 1.016 \times 10^5 \times 4.60 \times 10^{10} = 4.7 \times 10^{15} \mu m^3$$

Dividing this by the volume of a cell, 250,000 μm^3, we obtain 1.88×10^{10} or about 19 billion cells.

15-10 The area of the end of a plasmodesmata tube is

$$A = \pi \times 20^2 = 1256 \text{ nm}^2 = 1256 \text{ nm}^2 \times \frac{10^{-6} \mu m^2}{1 \text{ nm}^2} = 1.256 \times 10^{-3} \mu m^2$$

a. If the density is 25 per μm^2 they occupy $25 \times 1.256 \times 10^{-3} = 3.14 \times 10^{-2} \ \mu m^2$ per μm^2, or 3.14% of the area.

b. If the density is 0.2 per μm^2 they occupy $0.2 \times 1.256 \times 10^{-3} = 2.52 \times 10^{-4} \ \mu m^2$ per μm^2, or 0.025% of the area.

15-11   **a.** The inner radius is $5 - 0.3 = 4.7$ cm. The outer radius is 5.3 cm.

b. One-fourth the circumference of the inner circle is $(1/2)\pi r = (1/2)\pi(4.7) = 7.38$ cm. Similarly, the outer length is 8.33 cm.

c. The ratio is $7.38/8.33 = 0.89$.

d. For the inner length

$$\frac{7.38 \text{ cm}}{100 \ \mu m/\text{cell}} \times \frac{10^4 \ \mu m}{1 \text{ cm}} = 738 \text{ cells}$$

Similarly, for the outer length 833 cells.

d. Let $100 + x \ \mu m$ be the length of a cell on the outer arc. Because the average length is 100 μm, the length of a cell on the inner arc is $100 - x \ \mu m$. But from (c) we know the inner cell length is 0.89 times the outer cell length: $0.89(100 + x) = 100 - x$. Solving, we find x is approximately 6 μm. So the inner cell length is about 94 μm and the outer length is 106 μm.

CHAPTER 16

Long-Distance Transport

16-1   **a.** First convert the distance to meters: $1 \text{ mm} = 10^{-3} \text{ m}$. Then substitute in the formula.

$$t = \frac{(10^{-3})^2 \text{ m}^2 \text{ sec}}{2.4 \times 10^{-5} \text{ m}^2} = 0.042 \text{ sec}$$

b. Similarly,

$$t = \frac{(2 \times 10^{-3})^2 \text{ m}^2 \text{ sec}}{2.4 \times 10^{-5} \text{ m}^2} = 0.167 \text{ sec}$$

16-2   **a.** First convert the distance to meters: $50 \mu\text{m} = 5 \times 10^{-5}$ m. Then substitute in the formula.

$$t = \frac{(5 \times 10^{-5})^2 \text{ m}^2 \text{ sec}}{10^{-9} \text{ m}^2} = 2.5 \text{ sec}$$

b. Similarly,

$$t = \frac{1 \text{ m}^2 \text{ sec}}{10^{-9} \text{ m}^2} = 1 \times 10^9 \text{ sec}$$

Converting seconds to years, in this case a more comprehensible unit,

$$1 \times 10^9 \text{ sec} \times \frac{1 \text{ min}}{60 \text{ sec}} \times \frac{1 \text{ hr}}{60 \text{ min}} \times \frac{1 \text{ day}}{24 \text{ hr}} \times \frac{1 \text{ yr}}{365 \text{ days}} = 31.7 \text{ yr}$$

16-3   **a.** Note that two molecules of NH_3 are made during this reaction.
$4 \times 51.7 + 8 \times 7.3 = 265.2 \text{ kcal/mol}$.

b. $2 \times 51.7 + 3 \times 7.3 = 125.3 \text{ kcal/mol}$.

c. $265.2/125.3 = 2.12$ times as energy expensive.

d. $265.2 + 51.7 + 7.3 = 324.2 \text{ kcal/mol}$.

16-4

a. If you plot the points carefully and draw a smooth curve through them, you can read from the graph approximately

RH (%)	Ψ (MPa)
30	-365
40	-280
60	-160
80	-75

16-5

a. Using $P = -0.01$ MPa, we must solve for r in the equation

$$-0.01 \text{ MPa} = -2 \times \frac{7.28 \times 10^{-8} \text{ MPa} \cdot \text{m}}{r}$$

The result is $r = 1.456 \times 10^{-5}$ m $= 14.56 \mu$m.

b. Solve the same equation again for $P = -3$ MPa. The result is $r = 4.85 \times 10^{-8}$ m $= 0.0485 \ \mu$m.

16-6

a. The average of these two values for the water potential is

$$P = \frac{-0.3 + (-3.0)}{2} = -1.65 \text{ MPa}$$

Then solving the equation

$$-1.65 \text{ MPa} = -2 \times \frac{7.28 \times 10^{-8} \text{ MPa} \cdot \text{m}}{r}$$

for r, we obtain $r = 8.82 \times 10^{-8}$ m $= 0.088 \ \mu$m.

16-7 Assume the entire weight of grass comes from the water in the grass. Then 80% of 30 lb is 24 lb of vacuolar water from 1/3 acre. This converts to

$$24 \text{ lb} \times \frac{1 \text{ kg}}{2.205 \text{ lb}} = 10.88 \text{ kg}$$

or 10.88 liters of water.

a. For nitrate,

$$0.045 \text{ mol}/1^{-1} \times 10.88\ 1 = 0.49 \text{ mol}$$

Similarly, for phosphate we get 0.71 moles and for potassium 1.58 moles.

b. The molecular weight of nitrate is 62, thus 0.49 moles weigh $0.49 \times 62 = 30.38$ grams. This is from 1/3 acre, so 91 grams is lost per acre. Converting to pounds,

$$91 \text{ g} \times \frac{1 \text{ kg}}{1000 \text{ g}} \times \frac{1 \text{ lb}}{0.4536 \text{ kg}} = 0.2 \text{ lb}$$

of nitrate is lost. Similarly, the molecular weight of phosphate is 96 and 0.45 lb of phosphate is lost. The molecular weight of potassium is 39.1 and 0.41 lb of potassium is lost.

16-8   **a.** For leaf size 1 mm,

$$\frac{76 \text{ plastid genomes}}{\text{plastid}} \times \frac{10 \text{ plastids}}{\text{cell}} = \frac{760 \text{ plastid genomes}}{\text{cell}}$$

Similarly, for the other leaf sizes we compute 1500, 5510, and 5472.

c. Between 1 mm and 20 mm. The curve is steepest there.

d. Between 20 mm and 100 mm.

16-9   **a.** The increase of 482 grams in 792 hours means there was an average increase of 482/792 g/hr. It was moved through a cross-sectional area 20% of $18.6 \text{ mm}^2 = 3.72 \text{ mm}^2$. So $482/3.72/792 = 0.164 \text{ g/mm}^2/\text{hr}$ was the mass transfer rate.

b.

$$\frac{\dfrac{0.164 \text{ g}}{\text{mm}^2 \cdot \text{hr}}}{\dfrac{1.5 \text{ g}}{\text{cm}^3} \dfrac{1 \text{ cm}^3}{1000 \text{ mm}^3}} = 109.3 \text{ mm/hr}$$

or about 110 mm/hr.

c. 550 mm/hr.

d. 550 mm/hr = 55 cm/hr, so pumpkin is relatively slow.

CHAPTER 17

Plant Development

17-1 If we assign units so that the thickness of a palisade cell in the sunlight is 1, then the thickness of a leaf in the sunlight is 2. The thickness of a leaf in the shade is then 2/3 plus 1, or 5/3. The ratio $2/(5/3) = 6/5 = 1.2$, so the sun-grown leaf is 20% thicker.

17-2 **a.** Notice that $8 = 2^3$, so three S cycles of endoreduplication will multiply the amount of DNA by 8. Similarly $64 = 2^6$, $256 = 2^8$, $2048 = 2^{11}$, and $8192 = 2^{13}$, so the number of S cycles is respectively 6, 8, 11, and 13. These numbers may also be obtained by using logarithm base e or base 10. For example, solving

$$2^n = 2048$$

$$\ln 2^n = n \ln 2 = \ln 2048$$

$$n = \frac{\ln 2048}{\ln 2} = 11$$

b. However, solving

$$2^n = 24{,}576$$

$$n = \frac{\ln 24{,}576}{\ln 2} = 14.58$$

which is not an integer. The cells can have undergone at most 14 S cycles. Some gene amplification must have occurred.

17-3 **a.** It takes 13.5 days for the first flower to open. In that time the plant has made $13.5 \times 1.9 = 25$ flowers.

b. When the meristem stops making flowers, the most recently opened flower began 13.5 days previously, or 16.5 days after the beginning of flowering. During that time the plant made $16.5 \times 1.9 = 31$ flowers.

17-4   **a.**

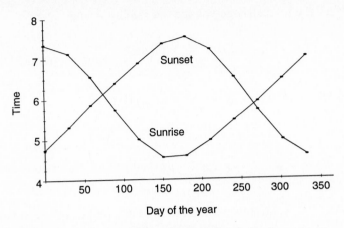

Sunrise and Sunset

b.

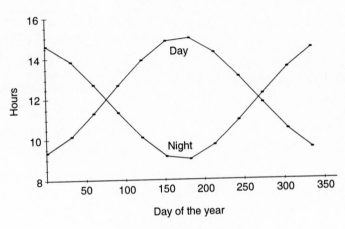

Day Length and Night Length

c. Dill is induced to flower about February 20 and spinach about April 9, when the days get long enough. Unless they were planted well earlier than that, they will flower as soon as they are mature. Cocklebur is induced to flower about August 7 when the days get short enough. The days are always shorter than 15.5 hours, so soybeans flower as soon as they are mature.

17-5   **a.** Because $(1132 - 816)/816 = 0.387$, there is a 39% percent increase.

b. If x is the side length, it is related to the volume by $V = x^3$ or $x = \sqrt[3]{V}$. So the side length increases from $\sqrt[3]{816} = 9.34 \mu m^3$ to $\sqrt[3]{1132} = 10.42 \ \mu m^3$. Because $(10.42 - 9.34)/9.34 = 0.116$, there is a 12% percent increase.

c. At 144 HAF the total storage volume per seed is

$$\frac{816\ \mu m^3}{cell} \times \frac{5400\ cells}{seed} = 4.4 \times 10^6 \mu m^3$$

At 216 HAF the total storage volume per seed is

$$\frac{1132\ \mu m^3}{cell} \times \frac{5400\ cells}{seed} = 6.1 \times 10^6 \mu m^3$$

17-6

a. In the first year, the circumference of the xylem is $C = \pi d = 20\pi = 62.83$ cm. The length of an individual cell is 8 μm or 8×10^{-4} cm. So this exposed face of the cambium layer is made up of

$$\frac{62.83}{8 \times 10^{-4}} = 7.85 \times 10^4 \text{ cells.}$$

Seventy-five percent of them, or 5.89×10^4 are fusiform initials and 25%, or 1.96×10^4 are ray initials.

b. For the second year we repeat the calculation using $d = 21$ cm and find there are now 8.25×10^4 cells of which 6.18×10^4 are fusiform initials and 2.06×10^4 are ray initials.

c. In the second year there were 1000 more ray initials, thus 1000 fusiform initials were converted.

17-7

a. We compute the radius in inches for which the volume of wood created during the next year is 6 board feet. Note that 8 ft = 96 in and 6 board feet = 864 in³. Thus, we must solve the following equation for r:

$$96\pi \left(r + \frac{1}{4}\right)^2 - 96\pi r^2 = 864$$

We find $r = 5.6$ inches.

The initial circumference $C = 2\pi r$, where $C = 6$ in, so the initial radius without the bark is

$$\frac{6}{2\pi} - 0.5 = 0.455 \text{ in}$$

So the radius has to increase from 0.455 in to 5.6 in at 0.25 in per year. This takes

$$\frac{5.6 - 0.455}{0.25} = 20.58 \text{ yr}$$

Rounding off, the tree starts the critical growth year after 21 years, so it is ready to harvest at 22 years.

17-8   **a.** There are $132,000 \times 0.36 = 47,520$ mesophyll cells and the same number of epidermis cells. There are $132,000 \times 0.28 = 36,960$ vascular cells.

The volume of a leaf is $30 \times (0.1) \text{ mm}^3 = 3 \text{ mm}^3$. The volume of the mesophyll cells is thus $0.605 \times 3 = 1.815 \text{ mm}^3$, the volume of the epidermis cells is thus $0.125 \times 3 = 0.375 \text{ mm}^3$, and the volume of the vascular cells is thus $0.015 \times 3 = 0.045 \text{ mm}^3$. Each mesophyll cell then has approximately the volume

$$\frac{1.815 \text{ mm}^3}{47,520} = 3.82 \times 10^{-5} \text{ mm}^3 = 38,200 \ \mu m^3$$

Similarly, the epidermis cells have a volume $7890 \ \mu m^3$ and the vascular cells $1220 \ \mu m^3$.

b. The three types of cells make up $60.5 + 12.5 + 1.5 = 74.5\%$ of the leaf. So 25.5% is air. The volume of air is $0.255 \times 3 = 0.77 \text{ mm}^3$.

17-9   **a.** The volume of a root hair is

$$V = \pi r^2 h = \pi \left(\frac{0.7}{2}\right)^2 (1.5) = 0.577 \text{ mm}^3$$

The volume of a vesicle is

$$V = \frac{4}{3} \pi r^3 = \frac{4}{3} \pi (0.05)^3 = 5.24 \times 10^{-4} \ \mu m^3$$

Converting the volume of a root hair to micrometers in order to have the same units,

$$0.577 \text{ mm}^3 \times \frac{10^9 \mu m^3}{1 \text{ mm}^3} = 5.77 \times 10^8 \ \mu m^3$$

Dividing,

$$\frac{5.77 \times 10^8 \ \mu m^3}{5.24 \times 10^{-4} \ \mu m^3} = 1.1 \times 10^{12} \text{ vesicles}$$

b. Using the area of the side walls plus one end,

$$SA = 2\pi r h + \pi r^2 = \pi r(2h + r) = \pi(0.35)(3 + 0.35) = 3.68 \text{ mm}^2$$

c. To go 1.5 mm $= 1.5 \times 10^3 \ \mu m$ at 100 μm per hour takes

$$t = \frac{1.5 \times 10^3}{100} = 15 \text{ hr}$$

d.

$$\frac{1.1 \times 10^{12} \text{ vesicles}}{15 \text{ hr}} = 7.3 \times 10^{10} \text{ vesicles per hour}$$

17-10   **a.** If the side length of the cube is x, the volume $V = x^3$ so $x = \sqrt[3]{V}$. Thus, the side length of a cell at hour 168 is $\sqrt[3]{951} = 9.83 \ \mu m$ and at hour 216 is $\sqrt[3]{1,132} = 10.42 \ \mu m$. The surface area is $SA = 6x^2$, so at hour 168 is 580 μm^2 and at hour 216 is 652 μm^2.

b. At hour 168, 38:951 reduces to the ratio 1:25. At hour 216, 13:1132 reduces to 1:87.

c. At hour 168, 1413:580 reduces to 1:0.41. At hour 216, 514:612 reduces to 1:1.19.

17-11   **a.** First convert milligrams per liter to moles per liter:

$$\frac{0.2 \text{ mg}}{1} \times \frac{10^{-3} \text{ g}}{1 \text{ mg}} = \frac{2 \times 10^{-4} \text{ g}}{1}$$

The mass of IAA is 180 g/mol.

$$\frac{2 \times 10^{-4} \text{ g/l}}{180 \text{ g/mol}} = 1.11 \times 10^{-6} \text{ mol/l} = 1.11 \times 10^{-6} \text{ M}$$

b. The volume of an apical meristem cell is $7^3 = 343 \ \mu m^3$. In liters,

$$343 \ \mu m^3 \times \frac{10^{-12} \text{ cm}^3}{1 \ \mu m^3} \times \frac{1 \text{ ml}}{1 \text{ cm}^3} \times \frac{10^{-3} \text{ l}}{1 \text{ ml}} = 3.43 \times 10^{-13} \text{ l}.$$

At 1×10^{-6} moles per liter, we have 3.43×10^{-19} moles. Multiplying by Avogadro's number, we see there are $(3.43 \times 10^{-7}) \times (6.02 \times 10^{23}) = 206,500$ molecules in a cell.

c. The distance across one cell is $15\ \mu\mathrm{m} = 0.0015$ cm. At 1 cm/hr it can therefore cross in one hour $1/0.0015 = 667$ cells.

17-12   **a.** If we assume a length of 25 cm or 250 mm, at 6.25 mm/hr it takes $250/6.25 = 40$ hours.

b. Almost all of the surface area comes from the inside and outside walls of the pollen tube, not the end. Thus the internal surface area, assuming a length of 25 cm, is

$$SA = 2\pi r h = 2\pi \times (5\ \mu\mathrm{m}) \times (25 \times 10^4\ \mu\mathrm{m}) = 7.854 \times 10^6\ \mu\mathrm{m}^2$$

The surface area of a vesicle of diameter 0.1 μm is

$$SA = 4\pi r^2 = 4\pi (0.05\ \mu\mathrm{m})^2 = 0.0314\ \mu\mathrm{m}^2$$

Thus $(7.854 \times 10^6)/0.0314 = 2.5 \times 10^8$ vesicles are required for the inside membrane, and the same number for the outside membrane, for a total of 5×10^8.

c. The cell wall volume is approximately the internal surface area times the thickness, or

$$V = (7.854 \times 10^6\ \mu\mathrm{m}^2) \times 0.02\ \mu\mathrm{m} = 1.57 \times 10^5\ \mu\mathrm{m}^3$$

The volume of one vesicle is

$$V = \frac{4}{3}\pi r^3 = \frac{4}{3}\pi (0.05\ \mu\mathrm{m})^3 = 5.236 \times 10^{-4}\ \mu\mathrm{m}^3$$

so the volume of 5×10^8 vesicles is $2.62 \times 10^5\ \mu\mathrm{m}^3$. The ratio of these volumes is $(1.57 \times 10^5)/(2.62 \times 10^5) = 0.599$, so 60% of the volume of the vesicles is used in the pollen tube wall.

d. Assuming the tip is 4 μm long and the silk is 25 cm long, the length of the vacuole is effectively still 250 mm. The radius is 5×10^{-3} mm. Hence the volume is

$$V = \pi r^2 h = \pi \times (5 \times 10^{-3})^2 \times 250 = 0.02\ \mathrm{mm}^3$$

CHAPTER 18

Comparative Physiology

18-1

a. If the goldfish could use *all* of the dissolved oxygen it would need to process

$$\frac{\frac{30 \text{ ml O}_2}{\text{kg} \cdot \text{hr}}}{\frac{9 \text{ ml O}_2}{\text{l}}} = 3.33 \text{ l/kg/hr}$$

Because it can use only 80% of the dissolved oxygen it must process $5/4 \times 3.33 = 4.17$ l/kg/hr.

b. Similarly $5/4 \times 285/5 = 71.3$ l/kg/hr.

c. $71.3/4.17 = 17.1$, or about 17 times as much.

18-2

a. Recall that a mole of O_2 occupies 22.4 liters or 22,400 ml. For the rat,

$$\frac{22,400 \text{ ml}}{(280 \text{ g}) \times (0.88 \text{ ml/g} \cdot \text{hr})} = 90.9 \text{ hr}$$

or 91 hr. Similarly, for the cat 17 hr, the dog 3.1 hr, the man 1.5 hr, and the cow 0.36 hr.

b. For the rat,

$$\frac{0.88 \text{ ml}}{\text{g} \cdot \text{hr}} \times 280 \text{ g} \times \frac{1 \text{ hr}}{60 \text{ min}} = 4.1 \text{ ml/min}$$

Similarly, for the cat 23 ml/min, the dog 120 ml/min, the man 250 ml/min, and the cow 1030 ml/min.

18-3

a. $w = k(2l)^3 = 8kl^3$. The weight multiplies by 8.

b. Let x be the unknown factor by which the length must increase to double the weight. We must solve for x:

$$2w = k(xl)^3$$

So $2w = x^3(kl^3) = x^3 w$, $x^3 = 2$ and $x = \sqrt[3]{2} = 1.26$.

c. Because $5 = 1.25 \times 4$, the weight would essentially double.

d. Using the data from the baby in the equation $w = kl^3$, $7 = k \times 20^3$ so $k = 8.75 \times 10^{-4}$. So the expected weight of an adult would be $w = (8.75 \times 10^{-4}) \times 70^3 = 300$ pounds.

18-4   **a.** The butterfly uses $100/0.7 = 140$ times as much, the fruit fly 12.5 times, the hummingbird 6.1 times, and the pigeon 13 times.

b. The closest is the pigeon.

18-5   **a.**
$$\frac{1 \text{ mm}}{1 \text{ sec}} = \frac{10,000 \text{ body lengths}}{1 \text{ hr}} \times \frac{1 \text{ hr}}{3600 \text{ sec}} = \frac{2.778 \text{ body lengths}}{1 \text{ sec}}$$

So 1 mm $= 2.778$ body lengths, or 1 body length $= 1/2.778 = 0.36$ mm.

b.
$$\frac{60,000 \text{ ft}}{1 \text{ hr}} \times \frac{1 \text{ mi}}{5280 \text{ ft}} = 11.36 \text{ mi/hr}$$

or about 11 mi/hr.

18-6   **a.** For copulant snails, $25.6 \times 28 = 717$. For noncopulant snails, $40.8 \times 28 = 1142$, for a total of about 1860 eggs.

b. $1142 - 717 = 425$.

c. $425/28 = 15.2$.

d. $(1142 - 717)/717 = 0.593$, so 59%.

18-7   **a.** A cheetah can run at the speed of (71 mi/hr) $\times$ (1 hr/3600 sec) $= 0.01972$ mi/sec. In 20 seconds it can cover $0.01972 \times 20 = 0.3944$ miles. A gazelle can run at (50 mi/hr) $\times$ (1 hr/3600 sec) $= 0.01389$ mi/sec, so in 20 seconds can go 0.2778 miles. The difference $0.3944 - 0.2778 = 0.1166$ miles or about 206 yards.

18-8   a. Assuming a 3/4 inch-thick insect, we must convert $3/8 = 0.375$ inches to meters:

$$d = 0.375 \text{ in} \times \frac{2.54 \text{ cm}}{1 \text{ in}} \times \frac{1 \text{ m}}{100 \text{ cm}} = 0.009525 \text{ m}$$

Substituting for d and D in the formula we obtain $t = 3.8$ seconds.

18-9   a. Each mole of ATP can create 6.02×10^{23} photons. Assume the flash takes 10^{14} photons. Then $(6.02 \times 10^{23} \text{ photons})/(10^{14} \text{ photons/flash}) = 6.02 \times 10^{9}$ flashes.

b. One flash requires 10^{14} molecules of ATP. So $(10^{14}$ molecules of ATP/flash$)/(36$ molecules ATP/molecule glucose$) = 2.8 \times 10^{12}$ molecules of glucose per flash.

18-10   a. 24.5 hr.   b. 72.2 cm.

c. $(0.122)(32.9) + (0.8)(0.3) + (0.984)(6.1) = 10.3$ g.

d. $32.9/33.6 = 0.98$ g/cm.   e. $32.9/3 = 11$ g/hr.   f. $6.1/20 = 0.31$ g/hr.

18-11   a. $50 \times 150 = 7500$ pounds.   b. $300 \times 150 = 45{,}000$ pounds.

c. $200 \times 6 = 1200$ feet.

d. The caterpillar eats an average of $86{,}000/24 = 3583$ times its weight in one day. So $3583 \times 7 = 25{,}081$ pounds of milk per day, or about 25,000 pints per day.

18-12   a. The ratio is $0.5/5000 = 0.0002$. So it uses 0.02% the normal amount of oxygen.

b. The ratio $(8000 - 5000)/5000 = 0.6$. So it overshoots the normal rate 60%.

c. Let us make the rough assumption that the hamster is at the alert level for 6 hours, and hibernating the rest of the time, for 90 hours. (In other words, that the curve in the figure can be replaced by a rectangular step.) Then the amount of O_2 consumed is $(5000 \text{ ml/kg/hr}) \times (0.05 \text{ kg}) \times (6 \text{ hr}) = 1500$ ml while alert, and $(0.5 \text{ ml/kg/hr}) \times (0.05 \text{ kg}) \times (90 \text{ hr}) = 2.25$ ml while hibernating, for a total use of 1502 ml O_2.

If alert the entire time it would use (5000 ml/kg/hr) × (0.05 kg) × (96 hr) = 24,000 ml. The ratio is 1502/24,000 = 0.0626, so it uses about 6% as much oxygen, and therefore expends about 6% the energy, during 4-day periods that are mainly spent hibernating.

18-13

a. The rumen weighs 500/7 = 71.43 pounds or 71.43 × 0.4536 = 32.4 kg. Assuming its contents have more or less the density of water, it contains 32.4 liters. Relative to bacteria, the protozoa in the rumen are negligible; there are 10^{11} bacteria per milliliter. Hence, the rumen contains $(3.24 \times 10^4$ ml$) \times (10^{11}$ bacteria/ml$) = 3.24 \times 10^{15}$ bacteria. At 10^{12} bacteria per gram, they weigh $(3.24 \times 10^{15}$ bacteria$)/(10^{12}$ bacteria/gram$) = 3.24 \times 10^3$ g = 3.24 kg or 3.24 × 2.205 = 7.14 or about 7 pounds.

b. The efficiency is 2/7.14 = 0.28 or 28%.

18-14

a. For norepinephrine, from the maximum at 1830 to the minimum the next day at 1230 is 18 hours. For serotonin, from the maximum at 1130 to the minimum at 2130 is 10 hours.

b. For norepinephrine, (0.29 − 0.22)/0.22 = 0.318, so the maximum is 30% larger. For serotonin, (0.73 − 0.62)/0.62 = 0.177, so 18%.

c. For norepinephrine, in the late morning. For serotonin, in the evening.

18-15   **a.**

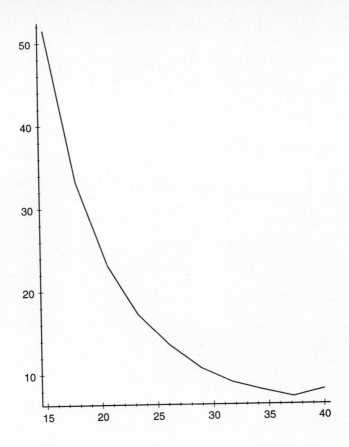

b. The graph is nonlinear.

c. The optimum temperature appears to be about 37°C.

d. To find the fraction of the distance that 25 is between the numbers 23.3 and 26.1, we compute $(25 - 23.3)/(26.1 - 23.3) = 0.61$. Hence, we take the same fraction of the distance between 17.2 and 13.5: $0.61 \times (17.2 - 13.5) = 2.257$. Subtracting this from 17.2 we obtain 14.9. Thus, we estimate the time needed at 25° to be 14.9 hours.

Similarly for 35°, $(35 - 34.4)/(37.2 - 34.4) = 0.21$. So $0.21 \times (8.1 - 7.2) = 0.189$. Hence, the time needed at 35° is approximately $8.1 - 0.189 = 7.9°$.

e. The first rate is the slope of the line joining the points (15.0, 51.5) and (25.0, 14.9), which is $(51.5 - 14.9)/(15 - 25) = -3.66$ hr/°C. The second rate is the slope of the line joining the points (25.0, 14.9) and (35.0, 7.9), which is $(4.9 - 7.9)/(25 - 35) = -0.70$ hr/°C.

18-16   **a.**

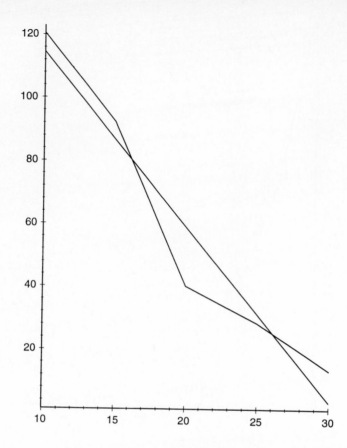

b. Computations give $\sum x_i = 100$, $\sum y_i = 295.2$, $\sum x_i^2 = 2250$, $\sum x_i y_i = 4515.5$, $m = -5.55$, and $b = 170$. So the line has equation $y = -5.55x + 170$.

c. The rate of decrease during the temperature interval $10°$ to $15°$ is $(120.5 - 92.4)/(10 - 15) = -5.62$. This is the closest to m of the four such slopes.

18-17   **a.** We make the assumption that the energy per gram of food is the same for both warm- and cold-blooded animals. Let x be the number of grams of food consumed by a homeotherm. Then $0.775x$ is assimilated and $(0.02)(0.775)x$ contributes to growth. Solving

$$(0.02)(0.775)x = 1$$

we obtain $x = 64.5$ g. Similarly, for poikilotherms the equation is

$$(0.44)(0.419)x = 1$$

and $x = 5.42$ g.

18-18   **a.** For lobsters: For cations Na^+ and K^+ we use the first equation. For Na,

$$E = 0.058 \log\left(\frac{465}{46}\right) = 0.0583 \text{ v}$$

Comparing this to the equilibrium state -0.07 v, we see this ion is out of equilibrium. Similarly for K, -0.085 v is somewhat less out of equilibrium. For anion Cl^- we use the second equation and obtain -0.056, also out of equilibrium.

The corresponding numbers for alga are -0.066 v, -0.178 v, and 0.098 v. Comparing these to -0.138 v, they too are all out of equilibrium.

b. For Na in lobsters, the inside concentration x that leads to equilibrium is the solution to the equation

$$0.058 \log\left(\frac{465}{x}\right) = -0.07$$

We obtain $x = 7540$ at equilibrium. The observed value 46 is $(7540 - 46)/7540 = 0.99$ or 99% less than the concentration at equilibrium.

For K in lobsters, $x = 162$. The observed value 292 is 80% over equilibrium. For Cl in lobsters, $x = 33$. The observed value 57 is 73% over equilibrium. For Na in alga, $x = 239$. The observed value 14 is 83% less than equilibrium. For K in alga, $x = 24$. The observed value 119 is 396% over equilibrium. For Cl in alga, $x = 0.0054$. The observed value 65 is 1,200,000% over equilibrium.

18-19   **a.** For human shoulder skin the average is 2.35, for rat liver 5.91, for rat epidermis 8.54 mitosing cells per 1000.

b. The fastest is rat epidermis. From the ratio $(8.54 - 2.35)/2.35 = 2.634$, we see it is 260% faster than human shoulder tissue.

c. For human shoulder skin, $(6.05 - 0.9)/0.9 = 5.72$, so at the fastest the cells are mitosing 570% faster than at the slowest time. For rat liver, $(13.1 - 2.1)/2.1 = 5.24$, so 520% faster. For rat epidermis, $(14.0 - 4.9)/4.9 = 1.86$, so 190% faster.

18-20   **a.** At 14.6°,

$$\frac{1741 \text{ kcal}}{mm^2 \cdot day} \times \frac{100 \text{ mm}^2}{1 \text{ cm}^2} = \frac{174,100 \text{ kcal}}{cm^2 \cdot day}$$

So

$$\frac{174,100 \text{ kcal}}{cm^2 \cdot day} \times \frac{1 \text{ cm}^2}{9.0 \text{ g}} = 19,344 \text{ kcal/g/day}$$

or 19.3 kcal/kg/day. The other four temperatures correspond in order to 11.5, 10.6, 9.77, and 11.2 kcal/kg/day.

b. The relationship is nonlinear for the mouse and linear for the reptile.

c. The slope of the graph for the mouse is largest from 30° to 35°.

d. At 26° the mouse produces about 10.4 kcal/kg/day and the reptile 2.3. Since $(10.4 - 2.3)/2.3 = 3.52$, the mouse generates about 350% more heat.

e. The slope of the reptile graph is $(3.2 - 1.2)/(32 - 18) = 0.143$. So for each extra degree in environmental temperature, 0.14 kcal/kg/day is produced.

CHAPTER 19

Human Physiology

19-1

a. Assume that the number of branchings is 23, so that the original tube ends in 2^{23} clusters of alveoli. Assume also that there are 400 million alveoli. Then there are $(4 \times 10^8)/2^{23} = 48$ alveoli per cluster.

19-2

a. Assume the blood flow to the heart increases to 3 times the amount; then it would provide 675 ml/min.

b. $225/5000 = 0.045$, so 4.5% of the blood flows to the heart during rest.

c. Assume the heart pumps 5 times faster during exercise; then $675/25,000 = 0.027$, so 2.7% of the blood flows to the heart during exercise.

19-3

a. For the resting athlete,

$$\frac{5 \text{ l/min}}{50 \text{ beats/min}} \times \frac{1000 \text{ ml}}{1 \text{ l}} = 100 \text{ ml/beat}$$

For the resting nonathlete,

$$\frac{5 \text{ l/min}}{72 \text{ beats/min}} \times \frac{1000 \text{ ml}}{1 \text{ l}} = 69 \text{ ml/beat}$$

For the exercising athlete,

$$\frac{30 \text{ l/min}}{185 \text{ beats/min}} \times \frac{1000 \text{ ml}}{1 \text{ l}} = 167 \text{ ml/beat}$$

For the exercising nonathlete,

$$\frac{22 \text{ l/min}}{195 \text{ beats/min}} \times \frac{1000 \text{ ml}}{1 \text{ l}} = 113 \text{ ml/beat}$$

b. At rest, $100/69 = 1.45$. Exercising, $167/113 = 1.48$.

19-4

a. Assuming the aveolus is a sphere, its surface area at radius 50 μm is

$$SA = 4\pi r^2 = 4\pi (50)^2 = 31,400 \ \mu m^2$$

The upper end of the range at radius 150 μm is 283,000 μm^2.

b. With average radius 100 μm, the surface area is 126,000 μm^2. There are two lungs with 3 × 10^8 alveoli each, for total area 2 × (3 × 10^8) × (1.26 × 10^5) = 7.56 × 10^{13} μm^2 or 75.6 m^2.

c. The surface area of the lungs is much larger. Because (75.6 − 2)/2 = 36.8, they are 3700% larger.

19-5   **a.** To move 5000 ml at 70 ml/beat takes 5000/70 = 71.4 beats. Since there are 72 beats per minute, it takes about 1 minute.

b. At rest, you move 70 ml/beat × 72 beats/min = 5040 ml/min. The marathon runner can move 7 times as much, or 35,280 ml/min. Assume it takes 190 beats to do this. Then 35,280 ml/min ÷ 190 beats/min = 186 ml/beat.

c. At 365 days/yr, 65 years = 23,725 days = 569,400 hours = 3.42 × 10^7 minutes. At 5040 ml/min, 1.72 × 10^{11} ml of blood is moved. This is 1.72 × 10^8 liters or (1.72 × 10^8) × 0.264 = 4.54 × 10^7 gallons.

d. The volume in the cans is 24 × 12 = 288 oz or 288 × 29.57 = 8516 ml. At 5040 ml/min it takes 1.7 minutes.

e. The volume in the pool is 10^6 gallons or 3.785 × 10^6 liters or 3.785 × 10^9 milliliters. At 5040 ml/min it takes 7.51 × 10^5 minutes, or 12,500 hours, or 521 days, or 1.43 years.

19-6   **a.** Assume the pancreas weighs 60 g and the endocrine portion is 1.5% of its weight. Then it weighs 60 × 0.015 = 0.9 g. The 60% that produces insulin weighs 0.9 × 0.6 = 0.54 g.

b.
$$\frac{0.54 \text{ g}}{70 \text{ kg}} \times \frac{1 \text{ kg}}{1000 \text{ g}} = 7.7 \times 10^{-6}$$

so 0.00077% of the weight is insulin-producing cells.

c. At 25 ng/kg body wt/min, a 70 kg person produces 70 × 25 = 1750 ng/min. Thus 0.54 g of cells produce 1750 ng/min. They produce their own weight in

$$\frac{0.54\text{g}}{1750 \text{ ng/min}} \times \frac{10^9 \text{ ng}}{1 \text{ g}} = 3.09 \times 10^5 \text{ min}$$

or 5.15 × 10^3 hr or 215 days.

19-7

a. At rest, a person moves 12 breaths/min × 0.5 l/breath = 6 l/min of air. During exercise, 48 breaths/min × 3 l/breath = 144 l/min.

b. The ratio $(144 - 6)/6 = 23$, so 2300% more air is moved during strenuous exercise.

c. $140,000 \text{ ft}^3 = 140,000 \times 12^3 \text{ in}^3 = 140,000 \times 12^3 \times 16.387 \text{ cm}^3 = 140,000 \times 12^3 \times$ 16.387 ml = $140 \times 12^3 \times 16.387$ l = 3,964,343 l. At 6 l/min it would take 660,724 min, or 11,012 hr, or 460 days.

19-8

a. A 70 kg person releases $70 \times 25 = 1750$ ng/min, or 1750×10^{-9} g/min. At 5808 g/mol, this is

$$\frac{1750 \times 10^{-9} \text{ g/min}}{5808 \text{ g/mol}} = 3.01 \times 10^{-10} \text{ mol/min}$$

So $(3.01 \times 10^{-10})(6.02 \times 10^{23}) = 1.8 \times 10^{14}$ molecules per minute are released.

b. Plasma volume is $0.55 \times 5 = 2.75$ l or 2750 ml. At 0.5 ng/ml, $0.5 \times 2750 = 1375$ ng of insulin are present during fasting.

$$\frac{1375 \times 10^{-9} \text{ g}}{5808 \text{ g/mol}} = 2.37 \times 10^{-10} \text{ mol}$$

or $(2.37 \times 10^{-10})(6.02 \times 10^{23}) = 1.4 \times 10^{14}$ molecules are present.

19-9

a. The ratio is $650/5000 = 0.13$, so 13%.

b. The ratio is $125/650 = 0.192$, so 19%.

c. 0.65 l/min × 60 min/hr × 24 hr/day = 936 l/day, or about 940 l/day.

d. Assuming a liter of blood plasma weighs one kilogram, 936 kg/70 kg = 13.37, so over 13 times as much.

e. Total glomerular filtrate in one day is 0.125 l/min × 60 min/hr × 24 hr/day = 180 l/day. Dividing by the number of glomeruli 2.4×10^6 we obtain 75 μl.

19-10   **a.** For T_3, we convert 30 μg to molecules:

$$\frac{30 \times 10^{-6} \text{ g}}{650 \text{ g/mol}} \times (6.02 \times 10^{23} \text{ molecules/mol}) = 2.78 \times 10^{16}$$

molecules of T_3. Similarly, there are 6.2×10^{16} molecules of T_4. So there are $(2.78 \times 10^{16}) + (6.2 \times 10^{16}) = 9 \times 10^{16}$ molecules of thyroxine.

Assuming the average, 16.5 mg cortisol per day, we find there are 2.87×10^{19} molecules of cortisol. Assuming 125 μg aldosterone, we obtain 2.1×10^{17} molecules of aldosterone.

b. There are 2.75 liters, or 27.5 dl of plasma, with 120 ng/dl T_3. So there are $(27.5)(120) = 3300$ ng of T_3. Converting to molecules,

$$\frac{3.3 \times 10^{-6} \text{ g}}{650 \text{ g/mol}} \times (6.02 \times 10^{23} \text{ molecules/mol}) = 3.06 \times 10^{15}$$

molecules of T_3, and similarly 1.70×10^{17} molecules of T_4. So there are 1.7×10^{17} molecules of thyroxine.

Assuming 60 ng/ml cortisol, we obtain similarly 2.9×10^{17} molecules of cortisol. Assuming 0.125 ng/ml aldosterone, we obtain 5.8×10^{14} molecules of aldosterone.

19-11   **a.** Convert to square feet:

$$\frac{(2 \times 10^6 \text{ cm}^2) \times (0.394^2 \text{ in}^2/\text{cm}^2)}{144 \text{ in}^2/\text{ft}^2} = 2200 \text{ ft}^2$$

b. The surface area of a tube of length 280 cm and circumference $2\pi r = 4\pi$ cm is $280 \times 4\pi = 3520$ cm^2. The ratio $(2,000,000 - 3520)/3520 = 567$, so 57,000% larger.

c. 280 minutes or 4 hours and 40 minutes.

19-12   **a.** Ten thousand in 10 days is 1000 per day or $1000/24 = 41\frac{2}{3}$ in one hour.

b. One taste bud is replaced in 1/1000 day; 2 are replaced in 1/500 day or 2.88 minutes.

19-13

a. For a resting heart, (70 ml/beat)(72 beats/min) = 5040 ml/min = 5.04 liters per minute.

b. Multiplying by the number of minutes in a day, we find 7258 liters per day.

c. For a stimulated heart, (210 ml/beat)(144 beats/min) = 30,240 ml/min = 30.2 liters per minute.

d. $(30.2 - 5.04)/5.04 = 5.0$, so 30.2 is 500% larger than 5.04.

e. Similarly 14 is 150% larger than 5.6.

19-14

a. 250 ml/m^2/min is $250 \times 60 = 15,000$ ml/m^2/hr during normal conditions, $15,000 \times 0.12 = 1800$ ml/m^2/hr in the cold, and 112,500 ml/m^2/hr during exercise.

b. Assume 7.5 lbs of water is lost. This weighs $7.5 \times 0.4536 = 3.4$ kg, so has volume 3.4 l. This is lost over 1.7 m^2 of skin area, so $3.4/1.7 = 2$ l or 2000 ml/m^2/hr.

19-15

a. In kilograms, 150 pounds is 68 kg. In centimeters, 5 feet is 152.5 cm and 9 inches is 22.86 cm, for total height 175 cm.

$$(0.007184)(68)^{0.425}(175)^{0.725} = 1.825 \text{ m}^2$$

b. Convert all units to square meters.

Sum of liver lobes	0.4 m^2
Skin	1.825 m^2
Sum of glomerular capillaries of the kidney	2 m^2
Lung	68 m^2
Small intestine	200 m^2
Bone cell surface area	4050 m^2

19-16

a. First we find the number of exhalations per day to be

$$\frac{12 \text{ exh}}{1 \text{ min}} \times \frac{60 \text{ min}}{1 \text{ hr}} \times \frac{24 \text{ hr}}{1 \text{ day}} = 17,280$$

Thus

$$\frac{500 \text{ ml/day}}{17{,}280 \text{ exh/day}} = 0.0289 \text{ ml/exh}$$

Converting this to units of weight, using the fact that water weighs 1000 grams per liter,

$$\frac{0.0289 \text{ ml}}{1 \text{ exh}} \times \frac{1 \text{ l}}{1000 \text{ ml}} \times \frac{1000 \text{ g}}{1 \text{ l}} = 0.029 \text{ g/exh}$$

b. Using the fact that water weighs 18 grams per mole, 0.029 grams is

$$\frac{0.029 \text{ g}}{18 \text{ g/mol}} = 1.61 \times 10^{-3} \text{ mol}$$

Finally, using Avogadro's number, there are
$(1.61 \times 10^{-3} \text{ mol}) \times (6.02 \times 10^{23} \text{ molecules/mol}) = 9.69 \times 10^{20}$ molecules per exhalation.

c. Using the answer to part (a), $(0.0289 \text{ g/exh}) \times (576 \text{ cal/g}) = 16.7$ calories per exhalation.

19-17   **a.** Of the exhaled air, 5.56%, or 0.0278 liters is CO_2 picked up from the lungs. Assuming standard temperature and pressure, one mole of gas occupies 22.4 liters. We compute the number of moles of CO_2 in 0.0278 l to be

$$\frac{0.0278 \text{ l}}{22.4 \text{ l/mol}} = 1.24 \times 10^{-3} \text{ mol}$$

The number of molecules is then $(1.24 \times 10^{-3}) \times (6.02 \times 10^{23}) = 7.46 \times 10^{20}$.
Seven percent of the molecules, or 5.22×10^{19} are carried in the plasma, 23% or 1.72×10^{20} are attached to hemoglobin, and 70% or 5.22×10^{20} are from HCO_3^-.

19-18   **a.** One mole of glucose requires 6 moles of O_2. Recall that one mole of gas occupies 22.4 liters. So 6 moles occupies $6 \times 22.4 = 134.4$ l, from which 686 kcal of energy is obtained. So the glucose broken down by one liter of O_2 yields $686/134.4 = 5.1$ kcal.

b. In the reaction, 1 mole of glucose yields 12 moles of water. As a gas it has volume $12 \times 22.4 = 268.8$ liters.

c. Water weighs 18 g/mol. So 12 moles of water weighs 216 g. Each gram of water occupies a milliliter, so as a liquid, 12 moles of water has volume 216 ml.

19-19   **a.** At 20 kcal/kg/day, the rate is 20/24 kcal/kg/hr. For a 70 kg student, this is 58.33 kcal/hr. Dividing, we find the oxygen requirement is

$$\frac{58.33 \text{ kcal/hr}}{4.8 \text{ kcal/l}} = 12.15 \text{ l/hr}$$

or about 12 l/hr.

b. Five times as much air is needed, or 60 liters.

c. Similarly for glucose,

$$\frac{58.33 \text{ kcal/hr}}{686 \text{ kcal/mol}} = 0.085 \text{ mol/hr}$$

Using the molecular weight of glucose we find $(0.085 \text{ mol/hr}) \times (180 \text{ g/mol}) = 15.3 \text{ g/hr}$.

d. Assume she runs for 2 hours, sleeps for 8 hours, is active for 7 hours, and is at the basal rate for 7 hours:

$$70 \left((2 \times 14.3) + \left(8 \times \frac{1.2}{24} \right) + \left(7 \times \frac{35}{24} \right) + \left(7 \times \frac{20}{24} \right) \right) = 3150 \text{ kcal}$$

19-20   **a.** There are 5×10^9 RBCs per milliliter. One pint contains 473 ml so $473 \times (5 \times 10^9) = 2.365 \times 10^{12}$, or about 2.4×10^{12} red blood cells.

b. At 15 g/dl, 4.73 dl weighs 70.95 g.

c. $(2.365 \times 10^{12} \text{ RBC}) \times (2.5 \times 10^8 \text{ molecules/RBC}) = 5.91 \times 10^{20}$ molecules of hemoglobin.

d. 5.91×10^{20} molecules of hemoglobin comprise

$$\frac{5.91 \times 10^{20} \text{ molecules}}{6.02 \times 10^{23} \text{ molecules/mol}} = 9.82 \times 10^{-4} \text{ mol}$$

of hemoglobin. Since 9.82×10^{-4} moles weighs 70.95 g, one mole weighs $(70.95 \text{ g})/(9.82 \times 10^{-4}) = 72{,}250$ g, so the molecular weight is about 72,000.

19-21   **a.** Assume that 30% of white blood cells are lymphocytes. Then there are $0.3 \times 7000 = 2100$ per mm^3. If 1000 of them are helper T cells, that leaves 1100 others per mm^3.

b. The ratio $1000/5{,}007{,}000 = 0.0002$, so 0.02% of the blood cells are helper T cells.

19-22   **a.** From $100/2 = 50$ to $100/1 = 100$ firings can occur.

b. The lengths of the intervals between firings are longest when action potentials are shortest, so let us assume the action potentials last 1 millisecond. Then during each second of the cramp, 300 milliseconds are taken up with firings and 700 milliseconds are spent resting. The duration of one rest period is $700/300 = 2.33$ ms.

19-23   **a.** Since milk is largely water, we assume that 1.5 l/day is the same as 1.5 kg/day. In pounds,

$$1.5 \text{ kg} \times 0.033 \times \frac{2.205 \text{ lb}}{1 \text{ kg}} = 0.109 \text{ lb/day}$$

Thus it takes $1/0.109 = 9.2$ days.

b. Similarly $1.5 \text{ kg} \times 0.002 = 0.003 \text{ kg/day} = 3 \text{ g/day}$.

19-24   **a.** Assume 1.5 mm^3 aqueous humor is made per minute. This is

$$\frac{1.5 \text{ mm}^3}{1 \text{ min}} \times \frac{60 \text{ min}}{1 \text{ hr}} \times \frac{24 \text{ hr}}{1 \text{ day}} = 2160 \text{ mm}^3/\text{day}$$

or 2160 μl per day. So the aqueous humor is replaced $2160/350 = 6.2$ times per day.

b. If 6 cm^3 produces 1.5 mm^3, 1 cm^3 produces $1.5/6 = 0.25$ mm^3 per minute or $0.25 \times 60 \times 24 = 360$ mm^3 per day.

c. 2160 μl = 2.16 ml. At 163 μmol/ml, 352 μmoles of sodium move through in a day. At 134 μmol/ml, 289 μmoles of chloride move through in a day.

19-25   **a.** Assuming an average nephron length 3 cm, their length is 3.6 million cm or 36,000 m. Converting to miles,

$$36,000 \text{ m} \times \frac{1.09 \text{ yd}}{1 \text{ m}} \times \frac{1 \text{ mi}}{1760 \text{ yd}} = 22 \text{ mi}$$

b. The circumference of a nephron is $c = 15\pi$ μm or $15\pi \times 10^{-6}$ m, so $(15\pi \times 10^{-6}) \times (3.6 \times 10^4) = 1.7 \text{ m}^2$.

19-26   **a.** Assuming the cilium is a cylinder and that the area of the end is negligible, the surface area is

$$SA = \pi dl = \pi(0.3)(65) = 61.3 \ \mu\text{m}^2$$

The total area of the cilia is then $61.3 \times 9 \times 10^8 = 552 \times 10^8 \ \mu\text{m}^2 = 552 \text{ cm}^2$, or about 550 cm^2.

b. The total olfactory surface area is 4.8 cm^2. The ratio is 552/4.8 = 115.

c. To be detectable, $(2.5 \times 10^{-10} \text{ mg/ml}) \times (500 \text{ ml}) = 1.25 \times 10^{-10}$ g must be present. The molecular weight of methyl mercaptan is 48.1 g/mol. So $(1.25 \times 10^{-10} \text{ g})/(48.1 \text{ g/mole}) = 2.60 \times 10^{-12}$ moles must be present. The number of molecules is $(2.60 \times 10^{-12}) \times (6.02 \times 10^{23}) = 1.57 \times 10^{12}$.

19-27   **a.** We must convert units and multiply by 5 dm^2:

$$\frac{2000 \text{ kcal}}{1 \text{ m}^2 \cdot \text{hr}} \times \frac{1 \text{ hr}}{60 \text{ min}} \times \frac{1000 \text{ cal}}{1 \text{ kcal}} \times \frac{1 \text{ m}^2}{100 \text{ dm}^2} \times 5 \text{ dm}^2 = 1670 \text{ cal/min}$$

b. At this rate in 108 seconds,

$$\frac{1670 \text{ cal}}{1 \text{ min}} \times \frac{1 \text{ min}}{60 \text{ sec}} \times 108 \text{ sec} = 3000 \text{ calories}$$

are lost.

19-28   **a.** While eating you produce $30 \times 2 = 60$ ml of saliva. While not eating you produce $(24 \times 60 - 30) \times 0.5 = 705$ ml, for a total of 765 ml of saliva per day.

b. At 100 units/ml, you produce 76,500 units of amylase per day.

c. At 1% concentration there is 1 g of starch per 100 ml of water, thus $1/20 = 0.05$ g per 5 ml water. Thus, one unit of amylase digests 0.05 g of starch in 1 minute.

d. In one day $= 60 \times 24 = 1440$ minutes, 1440 units of amylase could digest $0.05 \times 1440 = 72$ g of starch. As 76,500 is $76{,}500/1440 = 53.125$ times as much, you could digest $53.125 \times 72 = 3825$ g or 3.8 kg of starch in a day.

19-29   **a.** The totals are 73.6, 44.8, 25.8, 21.8, 19.2, 33.7.

b. The greatest difference is $73.6 - 19.2 = 54.4$.

c. The largest change is in G_1, $26 - 7.5 = 18.5$.

d. At $33°$, the time in the interphase is $26 + 22.4 + 12.2 = 60.6$. This is the maximum.

CHAPTER 20

Human Development and Reproduction

20-1   **a.** The total volume of sperm is 0.3 ml.

b. The volume of one sperm is $0.3/(3 \times 10^8) = 10^{-9}$ ml.

c. The fraction is $10^{-9}/3 = 3.3 \times 10^{-10}$, so one sperm occupies 0.000000033% of the volume.

d. Let the internal radius be r and the external radius be R. Then the volume is given by $V = \frac{4}{3}\pi R^3 - \frac{4}{3}\pi r^3$. We convert all distances to centimeters, so that $R = r + 0.3$ and $V = 50$ cm^3. We obtain the volume equation in cubic centimeters

$$50 = \frac{4}{3}\pi(R^3 - r^3)$$

$$\frac{50 \times 3}{4\pi} = (r + 0.3)^3 - r^3 = 0.9r^2 + 0.27r + 0.027$$

or

$$0.9r^2 + 0.27r - 11.91 = 0$$

Solving using the quadratic formula, we obtain

$$r = \frac{-0.27 \pm \sqrt{(0.27)^2 + 4(0.9)(11.91)}}{2(0.9)}$$

or $r = 3.5$ cm. The diameter is about 7 centimeters.

e. The surface area is given by $4\pi r^2 = 4\pi(3.5)^2 = 150$ cm^2.

20-2   **a.** One cm is 10,000 μm. Dividing, we find it takes 200 seconds or about $3\frac{1}{3}$ minutes.

b. Assume the fallopian tube is 11 cm long. Then the total distance is $1 + 7 + 10 = 18$ cm.

c. Eighteen cm at 200 cm/sec takes 3600 seconds or about an hour. (As sperm actually can get to the top of the fallopian tube in about 10 minutes, apparently some method other than swimming is being used.)

d. Assuming the egg survives 11 hours, and the sperm lives 3 days $= 72$ hours, the fraction is $(72 - 11)/11 = 5.55$. So the sperm lasts 555% longer than the egg.

e. Yes, it can swim the 18 cm in an hour and the egg lives at least 10 hours.

f. The speed of the swimmer in meters per hour is

$$\frac{50 \text{ m}}{21.91 \text{ sec}} \times \frac{3600 \text{ sec}}{1 \text{ hr}} = 8200 \text{ m/hr}$$

Dividing by the body length of the swimmer, 6 ft = 1.83 m, we find the speed is

$$\frac{8215 \text{ m/hr}}{1.83 \text{ m/body length}} = 4490 \text{ body lengths/hr}$$

The speed of the sperm is

$$\frac{50 \text{ } \mu\text{m}}{1 \text{ sec}} \times \frac{3600 \text{ sec}}{1 \text{ hr}} = 180{,}000 \text{ } \mu\text{m/hr}$$

This translates to

$$\frac{180{,}000 \text{ } \mu\text{m/hr}}{60 \text{ } \mu\text{m/body length}} = 3000 \text{ body lengths/hr}$$

The swimmer is faster.

20-3   **a.** The largest gain for boys is 16 pounds, in the first year. The next is 13.4 pounds, between ages 13 and 14. The largest gain for girls is 14.3 pounds, in the first year. The next is 12 pounds, between ages 11 and 12.

b. For boys, the percentage increase during the first year is the largest at 205%: $(23.8 - 7.8)/7.8 = 2.05$. The second year is the next largest at 22.7%. For girls, the percentage increase during the first year is the largest at 188%. The second year is the next largest at 26%.

20-4   **a.** 1,600,000 were lost, which is 80% of 2 million.

b. Her reproductive lifetime lasted 40 years at 365 days per year, or 14,600 days, plus 10 for leap years less $2 \times 270 = 540$ during gestation, for a total of 14,070 days. Dividing by 28, we find she released about 502 eggs.

c. At one egg for each meiosis event, and one egg for each 28 days, there is $1/28 = 0.036$ meiosis events per day.

d. At four sperm for each meiosis event, one-fourth of 100 million is 25 million meiosis events per day.

e. The ratio of meiosis events in the male to meisois events in the female is $(2.5 \times 10^7)/0.036 = 6.9 \times 10^8$.

20-5   **a.**

$$\frac{2 \text{ ng test.}}{1 \text{ mg t.t.}} \times \frac{1000 \text{ mg t.t.}}{1 \text{ g t.t.}} \times \frac{1 \text{ g test.}}{10^9 \text{ ng test.}} = 2 \times 10^{-6}$$

grams testosterone per gram testis tissue. As one gram testis tissue occupies about 1 milliliter, there are 2×10^{-3} grams testosterone per liter of testis tissue. At 300 grams per mole, the concentration is $(2 \times 10^{-3} \text{ g/l})/(300 \text{ g/mol}) = 6.67 \times 10^{-6} \text{ M} = 6.67 \ \mu\text{M}$, or about 6.7 μM.

b. The volume of a testis cell is

$$V = \frac{4}{3}\pi r^3 = \frac{4}{3}\pi \times (10 \ \mu\text{m})^3 = 4189 \ \mu\text{m}^3$$

Converting to liters,

$$(4189 \ \mu\text{m}^3) \times \frac{1 \text{ cm}^3}{10^{12} \ \mu\text{m}^3} \times \frac{1 \text{ ml}}{1 \text{ cm}^3} = 4.189 \times 10^{-9} \text{ ml} = 4.189 \times 10^{-12} \text{ l}$$

Converting to moles and finally molecules,
$(6.67 \ \mu\text{mol/l}) \times (4.189 \times 10^{-12} \text{ l}) = 2.8 \times 10^{-11} \ \mu\text{mol} = 2.8 \times 10^{-17} \text{ mol}$, or
$(2.8 \times 10^{-17}) \times (6.02 \times 10^{23}) = 1.7 \times 10^7$ molecules.

c. In 1 mg fetal testis tissue there are

$$2 \text{ ng} = \frac{2 \times 10^{-9} \text{ g}}{300 \text{ g/mol}} = 6.67 \times 10^{-12} \text{ mol}$$

or $(6.67 \times 10^{-12}) \times (6.02 \times 10^{23}) = 4.0 \times 10^{12}$ molecules of testosterone. So they can stimulate $(4.015 \times 10^{12}$ molecules$)/(10^4$ molecules/cell$) = 4.0 \times 10^8$ cells.

CHAPTER 21

Community Ecosystems and Ecology

21-1   **a.** The radius r at latitude θ (see the figure) is $r = 3964\cos\theta$.

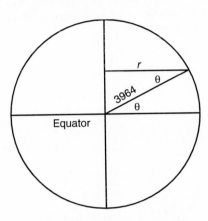

Cross Section of the Earth

The circumference at the equator is $2\pi \times 3964$ miles. The circumference at latitude θ is $2\pi \times 3964 \times \cos\theta$. Thus, the fraction of the distance traveled by a point at latitude θ as compared to a point at the equator is

$$\frac{2\pi \times 3964 \times \cos\theta}{2\pi \times 3964} = \cos\theta$$

Thus, if the speed at the equator is 425 m/sec, at latitude θ it is $425\cos\theta$ m/sec.
    For $\theta = 20°$, $425\cos(20°) = 399.4$ m/sec. Similarly $425\cos(40°) = 325.6$ m/sec and $425\cos(60°) = 212.5$ m/sec.

21-2   **a.** The male hunts for $0.448 \times 15 = 6.72$ hours and brings back $6.72 \times 0.46 = 3.09$ voles. Similarly, the female brings back 0.27 voles, for a total of 3.36 prey per day.

b. At 20.1 g/vole, 67.5 grams of food are gathered per day.

c. The mass of the prey gathered by the male is $3.09 \times 20.1 = 62.1$ g. The ratio is $62.1/181 = 0.343$, so the male gathers 34.3% of his weight in prey every day.

21-3

a. The circumference of a circle is 2π times the radius, so the radius of the earth is $24{,}906/2\pi = 3964$ miles. If we assume the albatross stays at 45° south latitude, we need to find the radius r of this circle. A cross section of the earth is shown in the figure.

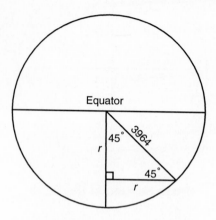

Cross Section of the Earth

Using the Pythagorean theorem with the 45-45-90 triangle, $r^2 + r^2 = 3964^2$ or $r = 3964/\sqrt{2} = 2803$ miles. So the circumference at 45° latitude is $2\pi \times 2803 = 17{,}600$ miles.

b. In a day, $17{,}600/365 = 48$ miles.

21-4

a. The forest area is $50^2 = 2500$ square miles, and the safe area is $40^2 = 1600$ square miles. Thus $1600/2500 = 0.64$ or 64% is safe nesting area.

b. The road has width 100 feet, or in miles,

$$100 \text{ ft} \times \frac{1 \text{ mi}}{5280 \text{ ft}} = 0.0189 \text{ mi}$$

Thus, the road has an area of $50 \times 0.0189 = 0.945$ square miles. The fraction lost is $0.945/2500 = 0.00038$, so the percent of forest lost is 0.038%.

c. The width of the safe area is now $40 - 0.0189 - 10$ miles, counting 5 miles on each side of the road. Now about 1200 square miles are safe. The proportion of safe nesting area lost is 400/1600 or 25%.

21-5   **a.** At site A, $(13.8 \text{ kJ/g}) \times (2.84 \text{ g/day}) = 39.2 \text{ kJ/day}$. At site B, 17.2 kJ/day. Thus, site A yields $39.2/17.2 = 2.28$ times as much energy per day.

b. At site A, $(39.2 \text{ kJ/day})/(682 \text{ bites/day}) = 0.0575 \text{ kJ/bite}$. At site B, 0.00566 kJ/bite. Thus site A yields $0.0575/0.00566 = 10.2$ times as much energy per bite.

c. At site A, $(39.2 \text{ kJ/day})/(23.2 \text{ min/day}) = 1.69 \text{ kJ/min}$. At site B, 0.211 kJ/min. Thus, site A yields $1.69/0.211 = 8.01$ times as much energy per minute feeding time.

21-6   **a.** We convert 5 kg/ha into lb/acre:

$$\frac{5 \text{ kg}}{1 \text{ ha}} \times \frac{1 \text{ ha}}{2.47 \text{ acres}} \times \frac{2.205 \text{ lb}}{1 \text{ kg}} = 4.46 \text{ lb/acre}$$

So with this pasture the 500-pound steer requires $(500 \text{ lb})/(4.46 \text{ lb/acre}) = 112$ acres. At 302 kg/ha only 1.85 acres are required.

21-7   **a.** First convert 7 lb/acre to g/ft^2:

$$\frac{7 \text{ lb}}{1 \text{ acre}} \times \frac{453.6 \text{ g}}{1 \text{ lb}} \times \frac{1 \text{ acre}}{43,560 \text{ ft}^2} = \frac{0.0729 \text{ g}}{1 \text{ ft}^2}$$

The molecular weight of $ZnSO_4 \cdot 7H_2O$ is 287.5.

$$\frac{\dfrac{0.0729 \text{ g}}{1 \text{ ft}^2}}{\dfrac{287.5 \text{ g}}{1 \text{ mol}}} \times \frac{6.02 \times 10^{23} \text{ molecules}}{1 \text{ mol}} = 1.53 \times 10^{20} \text{ molecules/ft}^2$$

b. The molecular weight of $CuSO_4 \cdot 5H_2O$ is 249.6. The equivalent calculation yields 1.76×10^{20} molecules per square foot.

c. There are 16 ounces in a pound, so $0.0729/7/16 = 6.5 \times 10^{-4} \text{ g/ft}^2$. The molecular weight of $Na_2MoO_4 \cdot 2H_2O$ is 241.9.

$$\frac{\dfrac{6.5 \times 10^{-4} \text{ g}}{1 \text{ ft}^2}}{\dfrac{241.9 \text{ g}}{1 \text{ mol}}} \times \frac{6.02 \times 10^{23} \text{ molecules}}{1 \text{ mol}} = 1.62 \times 10^{18} \text{ molecules/ft}^2$$

Chapter 21

21-8   **a.** The graph is the straight line through (55, 40) and (110, 90). The slope of this straight line is $m = (90 - 40)/(110 - 55) = 10/11$. Its equation is $y - 40 = (10/11)(x - 55)$ or the equivalent, $10x - 11y = 110$, where x is the number of rats and y is the percentage of pregnant pythons.

b. When $x = 20$, $y = 8\%$.

c. Using the equation, when $y = 0$, $x = 11$.

d. When $y = 100$, $x = 121$. Since $121 < 140$, it is possible.

e. Because the slope is 10/11, y increases by 10 when x increases by 11.

f. 5 kg + 15(55) g = 5.8 kg.

21-9   **a.** To calculate the amount of phosphorus in Nigerian soil in pounds per acre,

$$\frac{2 \text{ kg}}{1 \text{ ha}} \times \frac{1 \text{ ha}}{2.47 \text{ acres}} \times \frac{2.205 \text{ lb}}{1 \text{ kg}} = 1.79 \text{ lb/acre}$$

Similarly, 6.25 lb/acre K in Nigeria, 17.3 lb/acre P in Wisconsin, and 341 lb/acre K in Wisconsin.

b. Without soil amendments there is not enough of either P or K to generate a crop in Nigeria. There is enough of each in Wisconsin.

21-10   **a.** If the stalks were removed, the total N lost was $200 - 151 - 112 = -63$ kg/ha, P was gained at 6 kg/ha, and K was lost at -72 kg/ha.

b. All gained: N 49 kg/ha, P 24 kg/ha, K 63kg/ha.

21-11   **a.** Of the total water in rivers and lakes and groundwater, 1/61 of it is in rivers and lakes and 60/61 is in groundwater. So the amount in rivers and lakes is $0.0075 \times (3.59 \times 10^{17})/61 = 4.41 \times 10^{13}$ gallons.

b. In groundwater, there is 60 times this much, 2.65×10^{15} gallons.

c. In the atmosphere, $10^{-5} \times (3.59 \times 10^{17}) = 3.59 \times 10^{12}$ gallons.

d. Largest is groundwater, then water in rivers and lakes, U.S. daily rainfall, and atmospheric water.

21-12   **a.** $\dfrac{24 \times 10^6 \text{ km}^3}{361 \times 10^6 \text{ km}^2} = 0.066 \text{ km} = 66 \text{ m.}$

This is $66 \times 3.28 = 217$ feet.

21-13   **a.** The slope of the straight line through the points (1960, 320) and (1995, 360) is $m = (360 - 320)/(1995 - 1960) = 8/7$. Hence, the equation of the line is $y - 360 = 8/7(x - 1995)$. Substituting $x = 2030$ and solving for y, we find $y = 400$ ppm.

b. Exponential growth follows the law $N_t = N_0 e^{rt}$, where N_0 is the initial amount in 1960 (counted as year 0), r is the growth rate, t is the time in years, and N_t is the amount after t years. From the information given for 1995, $N_{35} = 360$, so

$$360 = 320 e^{35r}$$

Using natural logarithms to solve this equation, $\ln(360/320) = 35r$, so $r = 3.365 \times 10^{-3}$. Now we compute

$$N_{70} = 320 e^{(3.365 \times 10^{-3})(70)} = 405 \text{ ppm}$$

21-14   **a.** 56% of 115,000 = 64,400 kg is atmospheric N. Of this, 11% or 7084 kg enters the bay. 14% of 115,000 = 16,100 kg is from fertilizer, and of this 21% or 3381 kg enters the bay. 27% of 115,000 = 31,050 kg is from wastewater, and 35% or 10,870 kg enters the bay. The total amount is 21,335, so 21,340 kg.

b. From the air, 7084/21,335 = 0.332, so 33%. From fertilizer, 3381/21,335 = 0.158, so 16%. From wastewater, 10,870/21,335 = 0.509, so 51%.

c. 21,335/115,000 = 0.186, so 19%.

21-15   **a.** In 1910, $(290 \times 10^6$ acres$)/(90 \times 10^6$ people$) = 3.2$ acres/person.

b. In 1990, $(220 \times 10^6$ acres$)/(250 \times 10^6$ people$) = 0.88$ acres/person.

c. In 1910, 3.2 acres/person means $1/3.2 = 0.31$ persons/acre. In 1990, $1/0.88 = 1.14$ persons per acre. Thus, 100 acres would support 31 people in 1910 and 114 in 1990, or $114 - 31 = 83$ more people in 1990.

d. $(100 \times 10^6)/(30 \times 10^6) = 3\frac{1}{3}$ times more.

21-16   **a.** She consumed $365 \times 2 = 730$ pounds of food. The ratio $0.5/730 = 0.000685$. The percentage is 0.07%.

b. 10% of 730 is 73 pounds.

c. The net amount of weight gained was 12.5 pounds. The ratio $12.5/730 = 0.0171$, so the conversion percentage is 1.7%.

21-17   **a.** For confectioneries, assuming 10^3 cfu/g, in 75 g there would be 7.5×10^4 cfu. For cooked meats, 6×10^5, and for sandwiches and salads, 10^7.

b. For confectioneries and cooked meats, 100 times. For sandwiches and salads, 1000 times.

c. Assuming that one bacterium creates one cfu, for confectioneries, $(2.5 \times 10^6)/10^3 = 2.5 \times 10^3$ or 2500 g. For cooked meats, 250 g and for sandwiches and salads, 25 g.

21-18   **a.** The value of N for each of the six forests is 26, 30, 34, 27, 25, and 26. The values of H are 0.9157, 0.9290, 0.8362, 0.6039, 0.7621, 0.6811. The most diverse forest is thus the second Costa Rican forest, followed by 1, 3, 5, 6, 4.

21-19   **a.** The additional money that must be recovered in each of the two years is $7.50. If the increased yield is 1.6 bu, then $7.50/1.6 = $4.69 is the break-even price. If the increase is 3 bu, then $7.50/3 = $2.50.

b. If the increased yield is 2.5 bu, then $7.50/2.5 = $3.00 is the break-even price. If the increase is 5 bu, then $7.50/5 = $1.50.

c. 7.50/5.50 = 1.36 bu/acre. Since at least 1.6 bu/acre is expected, tillage is always worth the expense.

d. 7.50/1.95 = 3.85, slightly more than the average increased yield $(2.5 + 5)/2 = 3.75$ bu/acre. Tillage may not be worth the expense.

e. With interest of 10%, the price of tillage increases to $1.1 \times \$15 = \$16.50$, or $8.25 each year. If the increased yield is 1.6 bu, then $8.25/1.6 = $5.16 is the break-even price. If the increase is 3 bu, then $8.25/3 = $2.75. At $5.50 per bushel, tillage is still worth the expense.

21-20   **a.** One-third of 162.4×10^9 t, or 54.1×10^9 t is marine based. Two-thirds, 108.3×10^9 t, is land and freshwater based. One-third of this, or 36.1×10^9 t, is from tropical forests, distributed over 17×10^6 km^2. The productivity is

$$\frac{36.1 \times 10^9 \text{ t}}{17 \times 10^6 \text{ km}^2 \cdot \text{yr}} \times \frac{1 \text{ km}^2}{10^6 \text{ m}^2} \times \frac{10^6 \text{ g}}{1 \text{ t}} = 2120 \text{ g/m}^2/\text{yr}$$

b. Open ocean productivity, with units converted, is

$$\frac{127 \text{ g}}{1 \text{ m}^2 \cdot \text{yr}} \times \frac{10^6 \text{ m}^2}{1 \text{ km}^2} \times \frac{1 \text{ t}}{10^6 \text{ g}} = 127 \text{ t/km}^2/\text{yr}$$

Hence $(127 \text{ t/km}^2/\text{yr}) \times (332 \times 10^6 \text{ km}^2) = 42.2 \times 10^9$ t/yr comes from the open ocean. The ratio of this to all marine productivity is $42.2/54.1 = 0.780$, so 78% comes from the open ocean.

c. Tropical forest is $2120/127 = 17$ times more productive.

21-21   **a.** $0.72 \times 365 = 262.8$, so 260 kcal/m^2/yr.

b. Since the gross production of the tropical forest is 131 kcal/m^2/day, that of the temperate forest is $131/1.7 = 77.1$ kcal/m^2/day and the boreal forest is $131/2.7 = 48.5$ kcal/m^2/day. So for the tropical forest, the ratio is $0.72/131 = 0.0055$, or 0.55%. For the temperate forest, $0.72/77.1 = 0.0093$, or 0.93%. For the boreal forest, $0.72/48.5 = 0.0148$, or 1.5%.

21-22   **a.** First we need to convert to a common unit. Let us convert the diesel measurements to grams per mile. For carbon monoxide,

$$\frac{225 \text{ lb}}{1000 \text{ gal}} \times \frac{453.6 \text{ g}}{1 \text{ lb}} \div \frac{20 \text{ mi}}{1 \text{ gal}} = 225 \times 0.02268 = 5.1 \text{ g/mi}$$

Similarly, multiplying the others by 0.02268 we obtain 8.4 g/mi nitrogen oxides, 0.84 g/mi hydrocarbons, 0.07 g/mi aldehydes, 0.29 g/mi particulates, and 0.07 g/mi organic acids. The total is 14.8 g/mi for diesel and 97.7 g/mi for gasoline.

b. If diesel fuel gives x miles per gallon and we convert 370 lb/1000 gal to obtain the same number, 4.6 g/mi, that is emitted by gasoline, then we have the equation

$$\frac{370 \text{ lb}}{1000 \text{ gal}} \times \frac{453.6 \text{ g}}{1 \text{ lb}} \div \frac{x \text{ mi}}{1 \text{ gal}} = 4.6$$

that can be solved to yield $x = 36.5$ mi/gal.

c. Diesel releases 0.29 g/mi, compared to 0.6 from gasoline. So diesel is 2 times cleaner with respect to particulates.

21-23   **a.** If $x = 100$, $y = 5.46 + 7.76 = 13$.

b. If $y = 30$, $30 = 0.063x + 16.76$, so $x = 210$.

c. In summer, if $x = 200$, $y = (0.0546)(200) + 7.76 = 19$. In winter, $y = (0.063)(200) + 16.76 = 29$. So in winter the group is $(29 - 19)/19 = 0.53$ or 53% larger.

21-24   **a.** From $(117 - 45)/45 = 1.6$, we see the increase is 160%.

b. From $(73 \times 5 - 103)/103 = 2.54$, the increase is 254%.

c. First we compute the money earned on seeds only. If no bees,

$$\frac{35 \text{ seeds}}{1 \text{ yd}^2} \times \frac{4840 \text{ yd}^2}{1 \text{ acre}} \div \frac{275,000 \text{ seeds}}{1 \text{ lb}} = 0.616 \text{ lb/acre}$$

Similarly, if 1 colony, 0.792 lb; if 3 colonies, 1.267 lb; and if 5 colonies, 2.06 lb. So the income if no bees is $0.616 \times 1.50 = \$0.92$ per acre.

If 1 colony, $(0.792 \times 1.50) + (103 \times 2.65) = \$274.14$ per acre. If 3 colonies, $(1.267 \times 1.50) + (3 \times 92 \times 2.65) = \$733.30$, and if 5 colonies, $(2.06 \times 1.50) + (5 \times 73 \times 2.65) = \$970.34$.

d. If x is the price per colony, then we pay $5x$. We require that $5x + 0.1 \times 5x \leq 970.34$. Solving for x we obtain $x \leq \$176.42$.

21-25   **a.** The volume of a cone is $V = \frac{1}{3}\pi r^2 h$, where $h = 4000$ m and $r = 100$ m. We have only half a cone, so $V = 2.094 \times 10^7$ m^3. So the average weight is

$$\frac{0.01\mu g}{2.094 \times 10^7 \text{ m}^3} = 4.78 \times 10^{-10} \ \mu g/m^3$$

b. $0.01 \ \mu g = 10^{-8}$ g. Finding first moles and then molecules,

$$\frac{10^{-8} \text{ g}}{282 \text{ g/mol}} = 3.546 \times 10^{-11} \text{ mol}$$

$$(3.546 \times 10^{-11}) \text{ mol} \times (6.02 \times 10^{23}) \text{ molecules/mol} = 2.13 \times 10^{13} \text{ molecules}$$

or about 21 trillion molecules were released and dispersed.

c. Spread over 2.094×10^7 m^3 there are

$$\frac{2.13 \times 10^{13} \text{ molecules}}{2.094 \times 10^7 \text{ m}^3} = 1.0 \times 10^6 \text{ molecules/m}^3$$

or 1 million molecules per cubic meter.

CHAPTER 22

Population Ecology

22-1   **a.** $N = (22 \times 31)/11 = 62$.   **b.** The density is 6.2 voles per hectare.

22-2   **a.** The area of the base surface of the basidiocarp is the area of the ring between the two circles with diameters 7.5 cm and 2 cm. Let R and r be their radii.

$$A = \pi R^2 - \pi r^2 = \pi \times \left(\frac{7.5}{2}\right)^2 - \pi \times 1^2 = 41.04 \text{ cm}^2$$

At 40×10^6 spores per hour, we divide by 60 to get 6.67×10^5 spores per minute from the 41.04 cm² area, or $(6.67 \times 10^5)/41.04 = 16{,}000$ spores per square centimeter of basidiocarp area per minute.

b. Similarly, $(6.67 \times 10^5)/200 = 3300$ spores per square centimeter of gill surface area per minute.

c. At 40×10^6 spores per hour, 1.6×10^8 are released in 4 hours.

d. The volume of one spore is

$$V = \frac{4}{3}\pi \left(\frac{0.375}{2}\right)^2 \left(\frac{0.75}{2}\right) = 0.055 \ \mu\text{m}^3$$

So the entire volume is $1.6 \times 10^8 \times 0.055 = 8.8 \times 10^6 \ \mu\text{m}^3$.

22-3

Region	Pop. Size (millions)	Birth Rate (per thous.)	Births (millions)	Death Rate (per thous.)	Deaths (millions)	Growth Rate	Doubling Time (yr)
Africa	728	42	30.6	14	10.2	0.028	24.8
Asia	3458	25	86.5	8	27.7	0.017	40.8
Europe	727	12	8.7	11	8.0	0.001	693
Lat. Amer.	482	26	12.5	7	3.4	0.019	36.5
N. Amer.	293	16	4.7	9	2.6	0.007	99.0
Oceania	28	19	0.5	8	0.2	0.011	63.0
World	5716	25.1	143.5	9.1	52.1	0.016	43.3

22-4

a.

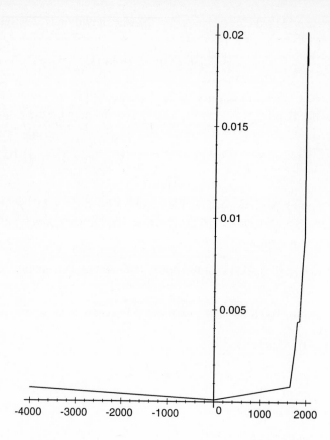

b. The exponential growth rates are 0.0007, 0.0001, 0.0009, 0.0029, 0.0044, 0.0044, 0.0071, 0.0090, 0.0181, 0.0202, 0.0184, 0.0173. Note that the growth rates themselves appeared to increase exponentially.

c. With $r = 0.0173$, $P_0 = 5285 \times 10^6$ and $T = 50$,

$$P_1 = 5.285 \times 10^9 \times e^{(50)(0.0173)} = 5.285 \times 10^9 \times 2.375 = 12.552 \times 10^9$$

or $12\frac{1}{2}$ billion people.

22-5

a. Solve the equation

$$3 = 2e^{0.0003t}$$

using the natural logarithm

$$\ln(3/2) = 0.0003t$$

to find $t = 1352$ days or about 3.7 years.

b. $N_{(3)(365)} = 2e^{(0.015)(3)(365)} = 27$ million rats.

c. $N_1 = 2e^{60} = 2.28 \times 10^{26}$ bacteria.

d. Solving $2e^{60} = 2e^{0.015t}$ we find $60 = 0.015t$, $t = 4000$ days or 11 years.

22-6   **a.** $1 - 1/\sigma \le 0$ for $\sigma \le 1$.

b. $\sigma = 1.4$.

c. $I = 1 - 1/1.4 = 0.286$, so 28.6% of the population is infectious.

22-7   **a.** $200,000 \times 0.78^7 = 35,130$.

b. $35,130/235,000,000 = (3.51 \times 10^4)/(2.35 \times 10^8) = 1.49 \times 10^{-4}$.

c. Using $p = 1.49 \times 10^{-4}$, $q = 0.9998$ and $1 - q^{7500} = 0.78$.

d. If $q = 0.999$, then $1 - q^{7500} = 0.9994$.

22-8   **a.** The graphs are displayed on page 298.

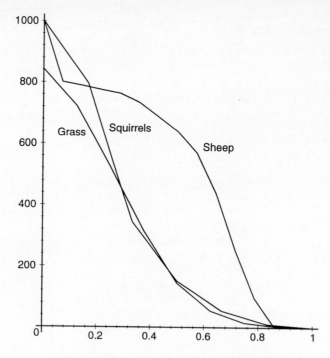

Sheep, Squirrel, and Grass Mortality

b. The curves for squirrels and grass are quite similar, but the curve for sheep is different.

c. For squirrels and grass, the mortality rate is fairly constant. For sheep, it is highest for the young and the old.

d. For sheep, half are still alive at $0.6 \times 14 = 8.4$ years. For squirrels, $0.28 \times 7 = 2.0$ years. For grass, $0.28 \times 24 = 6.7$ months.

22-9

a. The values obtained for the first regression line are

$n = 18, \sum x_i = 1.71 \times 10^2, \sum y_i = 3.87 \times 10^5, \sum x_i y_i = 4.88 \times 10^6, \sum x_i^2 = 2.11 \times 10^3, m = 2480, b = -2050$. For the second regression line,

$n = 7, \sum x_i = 1.47 \times 10^2, \sum y_i = 1.05 \times 10^5, \sum x_i y_i = 2.06 \times 10^6, \sum x_i^2 = 3.12 \times 10^3, m = -5580, b = 132,000$.

b. In 2000, for example, last year was 1999 and $x = 62$. So the expected number of cases would be $y = 2480 \times 62 - 2050 = 152,000$.

Math Review

Percentages

The word *percentage* comes from Latin and means "per hundred." To convert a fraction to a percentage, first write it with denominator 100.

Thus

$$\frac{1}{4} = \frac{25}{100}$$

so $\frac{1}{4}$ corresponds to 25 per hundred or 25%.

Alternatively we can divide, to write the fraction as a decimal, and then multiply by 100:

$$\frac{1}{4} = 0.25 = 25\%$$

The number 6 is 150% *of* the number 4 because

$$\frac{6}{4} = 1.5 = 150\%$$

We can also say the number 6 is 50% *larger than* the number 4 (in a language similar to that of marking up prices) because the additional amount $6 - 4 = 2$ is 50% of 4. In general if the number y is larger than the number x, to find the percentage by which it is larger than x we must do the computation

$$\frac{y - x}{x} \times 100$$

Scientific Notation

Often numbers arise in biology that are either very big or very small. In this case it is convenient to write them in *scientific notation*. A number in scientific notation has exactly one digit, other than 0, to the left of the decimal point.

The number

$$236000$$

in scientific notation is

$$2.36 \times 10^5$$

The number

$$0.00000078$$

is

$$7.8 \times 10^{-7}$$

The exponent of 10 tells how many places the decimal point has to be shifted. The positive number 5 shifts the decimal place in 2.36 five places to the right, while the negative number −7 shifts the decimal place in 7.8 seven places to the left.

The number

$$590.8 \times 10^4$$

needs to have the decimal point shifted two more places to the left in order to be in scientific notation. It becomes

$$5.908 \times 10^6$$

Previously, the decimal point needed to be moved four places to the right to obtain 5,908,000; now it has to be moved six.

The number

$$590.8 \times 10^{-4}$$

becomes

$$5.908 \times 10^{-2}$$

The decimal place was moved two places to the left, and still needs to be moved two more to obtain 0.05908.

Most calculators have some variation of the following procedure for entering numbers in scientific notation:

To enter 6.02×10^{23}:

press 6.02, press Exp, press 23

To enter 5.80×10^{-5}:

press 5.8, press Exp, press 5, press $+/-$

To enter -4.11×10^3:

press 4.11, press $+/-$, press Exp, press 3

Consult the calculator manual if necessary.

Appendix A

Significant Digits

No physical measurement can be absolutely precise. We can display Avogadro's number as

$$6.02 \times 10^{23}$$

in which case we are showing three significant digits (6, 0, and 2), or we may choose to display six if we know them:

$$6.02332 \times 10^{23}$$

The number of digits indicates the accuracy of a measurement. Similarly, constants such as π can be displayed with three significant digits

$$\pi = 3.14$$

or more

$$\pi = 3.14159265358979$$

If a number is written with a decimal point, then zeros on the right indicate significance. For example

$$5.80 \quad \text{and} \quad 700.$$

both indicate that three significant digits are known.

If a number is written without a decimal point, it may be known with total accuracy, such as in the statement that 1000 cm^3 equals 1 liter. It may also be ambiguous and require reasonable interpretation. A measurement given as 700 may intend either one, two, or three significant digits.

When rounding off, if the leftmost digit to be removed is less than 5, the preceding digit is not changed. If the leftmost digit to be removed is 5 or higher, the preceding digit is increased by 1. Thus if one digit is desired, 6.25 rounds to 6. (the leftmost digit to be removed is 2), but if two digits are desired then it rounds to 6.3 (the leftmost digit to be removed is 5).

When multiplying or dividing, the number of significant digits in the answer should be the smallest of the number of significant digits in the numbers going into the calculation. For example

$$\frac{10.6 - 4.5}{3.13} = 1.9488818$$

must be rounded to 1.9, since 4.5 has only two significant digits.

When adding or subtracting, the answer cannot have more digits to the right of the decimal point than any of the numbers going into the calculation. For example

$$4.61 + 11. + 8.2 = 23.81$$

must be rounded to 24., since 11. has no digits to the right of the decimal.

Multiplication and Division of Numbers in Scientific Notation

Example.   Find the product

$$(3.0 \times 10^8) \times (-2.0 \times 10^2)$$

To compute this by hand, we first regroup

$$(3.0 \times -2.0) \times (10^8 \times 10^2)$$

and multiply, obtaining

$$-6.0 \times 10^{10}$$

This example uses the fact that $10^8 \times 10^2 = 10^{10}$, which is a special case of one of the laws of exponents.

First Law of Exponents

$$a^m \times a^n = a^{m+n}$$

In the previous example, $a = 10$, $m = 8$, and $n = 2$. One way to think about this: To shift the decimal point eight places and then another two, is to shift it ten altogether.

The calculator finds the answer to the previous example immediately, without the intermediate regrouping step.

Example.   Find the product in scientific notation

$$564,000 \times 25,000$$

There are two reasonable ways to do this problem. One way is to first write each number in scientific notation

$$(5.64 \times 10^5) \times (2.5 \times 10^4)$$

and multiply, obtaining

$$1.41 \times 10^{10}$$

The other way is to enter the product 564,000 times 25,000 directly into the calculator, and then convert the answer to scientific notation. Since this number is so large, the step of converting the answer to scientific notation may be done automatically by the calculator.

The number of significant digits that appear in the answer to a calculation is never larger than the minimum of the numbers of significant digits in the input.

Example. Find the product

$$(6.9 \times 10^3) \times (7.85 \times 10^4)$$

Using a calculator, the answer is

$$5.4165 \times 10^8$$

However, 6.9 has only two significant digits and 7.85 has three. The answer is only accurate to the lesser of these two numbers, and therefore is

$$5.4 \times 10^8$$

The second law of exponents handles the case in which the exponent is zero.

Second Law of Exponents

$$a^0 = 1$$

for any number a (other than zero).

Example. Find the product

$$13000 \times 0.00040$$

This equals

$$(1.3 \times 10^4) \times (4.0 \times 10^{-4}) = (1.3 \times 4.0) \times (10^4 \times 10^{-4})$$

or

$$5.2 \times 10^0 = 5.2$$

The third law of exponents is needed for division.

Third Law of Exponents

$$\frac{a^m}{a^n} = a^{m-n}$$

Example.   Divide 176,000 by 0.022. We write

$$\frac{1.76 \times 10^5}{2.2 \times 10^{-2}} = 0.8 \times 10^{5-(-2)} = 0.8 \times 10^7 = 8.0 \times 10^6$$

Example.   Divide 0.001852 by 17.3; in other words, compute

$$\frac{1.852 \times 10^{-3}}{1.73 \times 10^1}$$

The result from the calculator is

$$1.0705202 \times 10^{-4}$$

Because we only know three signifiant digits of 17.3, we must round the answer to

$$1.07 \times 10^{-4}$$

In the course of a long computation it is better to save the rounding off until the last step, so as to avoid compounding round-off error. Also be sure to use enough digits of a constant such as π so as to be able to maintain as many significant figures as the measured data allows. If there is a key on the calculator for π, use it to get as many digits of accuracy here as possible.

Example.   Compute the volume of the sphere of radius $r = 1.36$ mm. The formula for the volume of a sphere is

$$r = \frac{4}{3}\pi r^3$$

Using the calculator key for π and doing the computation without intermediate rounding yields

$$10.536717$$

which must finally be rounded to

$$10.5 \, \text{mm}^3$$

Intermediate rounding after every multiplication and division leads to the answer

$$10.6 \, \text{mm}^3$$

Addition and Subtraction of Numbers in Scientific Notation

Addition and subtraction of numbers with the same exponent of 10 is done using the distributive law.

Example.   Add 4.56×10^3 and 5.90×10^3.

$$(4.56 \times 10^3) + (5.90 \times 10^3) = (4.56 + 5.90) \times 10^3$$

equals

$$10.46 \times 10^3 = 1.046 \times 10^4$$

which must be rounded to

$$1.05 \times 10^4$$

Example.   Subtract 8.42×10^{-2} from 7.81×10^{-2}.

$$(7.81 \times 10^{-2}) - (8.42 \times 10^{-2}) = -0.61 \times 10^{-2}$$

or

$$-6.10 \times 10^{-3}$$

If the exponents are different, one of the numbers must be rewritten so that the exponents agree before the addition can be done.

Example.   Add 6.84×10^8 and 5.00×10^6. If 6.84×10^8 is converted so that the exponent is 6,

$$6.84 \times 10^8 = 684 \times 10^6$$

we obtain

$$689 \times 10^6 = 6.89 \times 10^8$$

This could also have been done by converting 5.00×10^6 to 0.005×10^8 before adding. This conversion of exponents is done automatically by the calculator.

If one of the numbers being added or subtracted is very small compared to the other, it may be essentially insignificant.

Example.   Add 5.66×10^{-2} to 7.45×10^2. The calculator yields the answer

$$7.450566 \times 10^2$$

which must be rounded to

$$7.45 \times 10^2$$

Thus 5.66×10^{-2} is insignificant.

Conversion of Units

Frequently, quantities given in one unit of measurement must be converted to another during the course of a calculation. One way to do this is to multiply by the appropriate conversion factor taken from the table.

Example.   Convert 36.79 inches to centimeters. Using 1 in = 2.54 cm,

$$36.79 \, \text{in} = 36.79 \, \text{in} \times 2.54 \, \text{cm/in} = 93.4 \, \text{cm}$$

Another way to convert units consists in multiplying the number by 1, suitably expressed to allow cancellation of units.

Example.   Convert 60 miles per hour to feet per second.
Using the facts that there are 5280 feet in a mile and 3600 seconds in an hour,

$$60 \, \frac{\text{mi}}{\text{hr}} \times \frac{5280 \, \text{ft}}{1 \, \text{mi}} \times \frac{1 \, \text{hr}}{3600 \, \text{sec}} = \frac{60 \times 5280}{3600} \, \frac{\text{ft}}{\text{sec}} = 88 \, \text{ft/sec}$$

If a unit appears to a power higher than 1, the conversion factor will need to appear more than once also.

Example.   Convert 6,800,000 kilograms per cubic meter to grams per cubic millimeter.

$$\frac{6,800,000 \, \text{kg}}{\text{m}^3} \times \frac{1 \, \text{m}^3}{1000^3 \, \text{mm}^3} \times \frac{1000 \, \text{g}}{1 \, \text{kg}} = \frac{6,800,000 \, \text{g}}{1000^2 \, \text{mm}^3} = 6.8 \, \text{g/mm}^3$$

Logarithms

Any positive number x can be expressed as a power of another positive number $b > 1$ by using the correct exponent y:

$$x = b^y$$

This number y is called the *logarithm* of x with *base b*. The logarithm with base b of x is written

$$y = \log_b x$$

The two most frequently used bases for logarithms are 10 and the irrational number e. To three places,

$$e = 2.718$$

The logarithm with base 10 of x is written $\log x$ and the logarithm base e of x is written $\ln x$. The relationship between these two logarithms is

$$\ln x = 2.3025851 \log x$$

Logarithms base 10 and base e can be found by using the keys log and ln on the calculator. For example,

$$\log 2.569 = 0.4097641$$

$$\log 0.0032 = -2.49485$$

and

$$\ln 10 = 2.3025851$$

These statements are equivalent to the statements

$$10^{0.4097641} = 2.569$$

$$10^{-2.49485} = 0.0032$$

and

$$e^{2.3025851} = 10$$

By the second law of exponents, $y = 0$ corresponds to $x = 1$ with either base. The numbers $y > 0$ correspond to $x > 1$ and the numbers $y < 0$ correspond to $0 < x < 1$. The graphs of $y = \log x$ and $y = \ln x$ are shown in Figure 1.

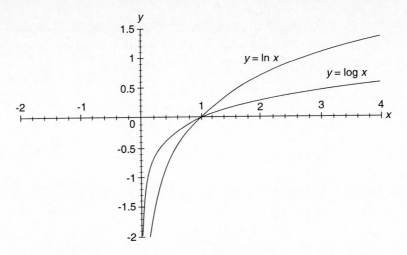

Figure 1. Logarithms

Logarithms are used in biology and chemistry to define the pH of a solution. If the hydrogen ion concentration $[H^+]$ in moles per liter is 10^{-7}, the pH of the solution is defined to be 7. Thus

$$pH = -\log_{10}[H^+].$$

Conversely, if the pH of a solution is 4, then the $[H^+]$ concentration is 10^{-4}.

Corresponding to the three laws of exponents are the three laws of logarithms.

Laws of Logarithms

$$\log(mn) = \log(m) + \log(n)$$

$$\log(1) = 0$$

$$\log\left(\frac{m}{n}\right) = \log(m) - \log(n).$$

Similar formulas hold in base e.

Exponential Growth and Decay

A quantity $A(t)$ that obeys the law

$$A(t) = A_0 e^{rt}$$

for a positive number r is said to be undergoing *exponential growth*. A quantity that obeys the law

$$A(t) = A_0 e^{-rt}$$

is undergoing *exponential decay*; again r is positive. In both cases, because $A(0) = A_0$, the quantity A_0 is the *initial amount*; the amount at time $t = 0$. The constant r is called the *growth rate* or *decay rate*. Figure 2 shows several graphs of exponential growth and decay.

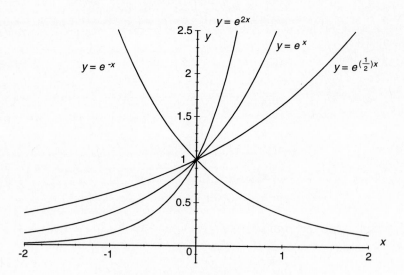

Figure 2. Some Graphs of $y = e^{rx}$

When a quantity is growing exponentially, there is a time, called the *doubling time t_2*, when the amount present is twice the initial amount A_0:

$$2A_0 = A_0 e^{rt}$$

Solving this equation for t, we find the doubling time is

$$t_2 = \frac{\ln 2}{r}$$

When a quantity is decaying exponentially, there is a time, called the *half-life $t_{\frac{1}{2}}$*, when the amount present is one-half the initial amount A_0:

$$\frac{1}{2}A_0 = A_0 e^{-rt}$$

Again solving for t, we obtain the formula for the half-life

$$t_{1/2} = \frac{\ln 2}{r}$$

Unit Conversion Factors

Unit and Abbreviation	Metric Equivalent	Metric-to-English Conversion Factor	English-to-Metric Conversion Factor
		LENGTH	
1 kilometer (km)	$= 1000\ (10^3)$ meters	1 km $= 0.62$ mile	1 mile $= 1.61$ km
1 meter (m)	$= 100\ (10^2)$ centimeters	1 m $= 1.09$ yards	1 yard $= 0.914$ m
	$= 1000$ millimeters	1 m $= 3.28$ feet	1 foot $= 0.305$ m
		1 m $= 39.37$ inches	
1 centimeter (cm)	$= 0.01\ (10^{-2})$ meter	1 cm $= 0.394$ inch	1 foot $= 30.5$ cm
			1 inch $= 2.54$ cm
1 millimeter (mm)	$= 0.001\ (10^{-3})$ meter	1 mm $= 0.039$ inch	
1 micrometer (μm)	$= 10^{-6}$ meter		
	$= 10^{-4}$ cm		
	$= 10^{-3}$ mm		
1 nanometer (nm)	$= 10^{-9}$ meter		
	$= 10^{-3}\ \mu$m		
1 angstrom (Å)	$= 10^{-10}$ meter		
	$= 10^{-4}\ \mu$m		
		AREA	
1 hectare (ha)	$= 10,000$ sq. meters	1 ha $= 2.47$ acres	1 acre $= 0.405$ ha
			1 acre $= 43,560$ sq. feet
1 sq. meter (m^2)	$= 10,000$ sq. centimeters	1 m$^2 = 1.196$ sq. yards	1 sq. yard $= 0.8361$ m^2
		1 m$^2 = 10.764$ sq. feet	1 sq. foot $= 0.0929$ m^2
1 sq. centimeter (cm^2)	$= 100$ sq. millimeters	1 cm$^2 = 0.155$ sq. inch	1 sq. inch $= 6.4516$ cm^2
		VOLUME (SOLIDS)	
1 cu. meter (m^3)	$= 10^6$ cu. centimeters	1 m$^3 = 1.308$ cu. yards	1 cu. yard $= 0.7646$ m^3
1 cu. centimeter (cm^3)	$= 10^{-6}$ cu. meter	1 cm$^3 = 0.061$ cu. inch	1 cu. inch $= 16.387$ cm^3
1 cu. millimeter (mm^3)	$= 10^{-9}$ cu. meter		
	$= 10^{-3}$ cu. centimeter		
1 cu. micrometer (μm^3)	$= 10^{-18}$ cu. m		
	$= 10^{-12}$ cu. cm		
	$= 10^{-9}$ cu. mm		

Unit and Abbreviation	Metric Equivalent	Metric-to-English Conversion Factor	English-to-Metric Conversion Factor

VOLUME (LIQUIDS AND GASES)

Unit and Abbreviation	Metric Equivalent	Metric-to-English Conversion Factor	English-to-Metric Conversion Factor
1 kiloliter (kl)	= 1000 liters	1 kl = 264.17 gallons	1 gallon = 3.785 l
1 liter (l)	= 1000 milliliters	1 l = 0.264 gallons	1 quart = 0.946 l
	= 10^{15} cu. micrometers	1 l = 1.057 quarts	
1 deciliter (dl)	= 10^{-1} liter		
1 milliliter (ml)	= 10^{-3} liter	1 ml = 0.034 fluid ounce	1 quart = 946 ml
	= 1 cu. centimeter	1 ml = $\frac{1}{4}$ teaspoon (approx.)	1 pint = 473 ml
			1 fluid ounce = 29.57 ml
1 microliter (μl)	= 10^{-6} liter		
	= 10^{-3} milliliter		
	= 1 cu. millimeter		

MASS

Unit and Abbreviation	Metric Equivalent	Metric-to-English Conversion Factor	English-to-Metric Conversion Factor
1 metric ton (t)	= 1000 kilograms	1 t = 1.103 tons	1 ton = 0.907 t
1 kilogram (kg)	= 1000 grams	1 kg = 2.205 pounds	1 pound = 0.4536 kg
			1 pound = 16 ounces
1 gram (g)	= 1000 milligrams	1 g = 0.0353 ounce	1 ounce = 28.35 g
1 milligram (mg)	= 10^{-3} gram		
1 microgram (μg)	= 10^{-6} gram		
1 nanogram (ng)	= 10^{-9} gram		

FORCE

Unit and Abbreviation	Metric Equivalent	Metric-to-English Conversion Factor	English-to-Metric Conversion Factor
1 newton (N)		1 N = 0.2248 pounds	1 pound = 4.448 N

ENERGY

Unit and Abbreviation	Metric Equivalent	Metric-to-English Conversion Factor	English-to-Metric Conversion Factor
1 joule (J)	= 1 m·N		
	= 2.389×10^{-4} kcal		
1 kcal	= 4186 J		

PRESSURE

Unit and Abbreviation	Metric Equivalent	Metric-to-English Conversion Factor	English-to-Metric Conversion Factor
1 megapascal (MPa)	= 10^6 pascals	1 MPa = 10 atmospheres (approx.)	
1 pascal (Pa)	= 1 newton/sq. meter	1 N = 1.45×10^{-4} lb/in^2	1 lb/in^2 = 6.895×10^3 Pa

TIME

Unit and Abbreviation	Metric Equivalent	Metric-to-English Conversion Factor	English-to-Metric Conversion Factor
1 second (sec)	= $\frac{1}{60}$ minute		
1 millisecond (ms)	= 10^{-3} second		
1 microsecond (μs)	= 10^{-6} second		
1 nanosecond (ns)	= 10^{-9} second		
1 picosecond (ps)	= 10^{-12} second		

Appendix B

Unit and Abbreviation	Metric Equivalent	Metric-to-English Conversion Factor	English-to-Metric Conversion Factor

MOLE–WEIGHT RELATIONSHIP

1 mole (mol)	= Avogadro's number of molecules or photons
1 millimole (mmole)	= 10^{-3} moles
1 kilo mol. wt. (kmwt)	= 10^{3} mwt
1 molecular weight (mwt)	= 1 dalton
1 dalton (d)	= 1 mwt

CONCENTRATION

1 molar (M)	= 1 mole per liter

TEMPERATURE

Unit and Abbreviation	Metric Equivalent	Metric-to-English Conversion Factor	English-to-Metric Conversion Factor
Degree Celsius (°C)		$°F = \frac{9}{5}°C + 32$	$°C = \frac{5}{9}(°F - 32)$
Degree Kelvin (K)	$0\,K = -273°C$		

Physical Constants

$\pi = 3.1416$

Avogadro's number $= 6.02 \times 10^{23}$

Planck's constant: $\hbar = 1.583 \times 10^{-34}$ calorie seconds

Speed of light: $c = 2.998 \times 10^8$ meters per second

One mole of gas at STP occupies 22.4 liters

The mass of 1 liter of water is 1 kilogram

Geometry Formulas

Diameter of a circle or sphere of radius r: $d=2r$

Circumference of a circle of radius r: $C=2\pi r$

Area of a circle of radius r: $A=\pi r^2$

Surface area of a sphere of radius r: $A=4\pi r^2$

Volume of a sphere of radius r: $\frac{4}{3}\pi r^3$

Surface area of a cylinder of height h and radius r: $A=2\pi r^2 + 2\pi rh$

Volume of a cylinder of height h and radius r: $V=\pi r^2 h$

Volume of a cone of height h and radius r: $V=\frac{1}{3}\pi r^2 h$

Essential Elements

Name	Symbol	Atomic number	Atomic weight
hydrogen	H	1	1.0
boron	B	5	10.8
carbon	C	6	12.0
nitrogen	N	7	14.0
oxygen	O	8	16.0
sodium	Na	11	23.0
magnesium	Mg	12	24.3
phosphorus	P	15	31.0
sulfur	S	16	32.1
chlorine	Cl	17	35.5
potassium	K	19	39.1
calcium	Ca	20	40.1
manganese	Mn	25	54.9
iron	Fe	26	55.8
cobalt	Co	27	58.9
copper	Cu	29	63.5
zinc	Zn	30	65.4
molybdenum	Mo	42	95.9

Chi-Square Distribution

Degrees of freedom	Area to the Right of the Critical Value									
	0.995	0.99	0.975	0.95	0.90	0.10	0.05	0.025	0.01	0.005
1	—	—	0.001	0.004	0.016	2.706	3.841	5.024	6.635	7.879
2	0.010	0.020	0.051	0.103	0.211	4.605	5.991	7.378	9.210	10.597
3	0.072	0.115	0.216	0.352	0.584	6.251	7.815	9.348	11.345	12.838
4	0.207	0.297	0.484	0.711	1.064	7.779	9.488	11.143	13.277	14.860
5	0.412	0.554	0.831	1.145	1.610	9.236	11.071	12.833	15.086	16.750
6	0.676	0.872	1.237	1.635	2.204	10.645	12.592	14.449	16.812	18.548
7	0.989	1.239	1.690	2.167	2.833	12.017	14.067	16.013	18.475	20.278
8	1.344	1.646	2.180	2.733	3.490	13.362	15.507	17.535	20.090	21.955
9	1.735	2.088	2.700	3.325	4.168	14.684	16.919	19.023	21.666	23.589
10	2.156	2.558	3.247	3.940	4.865	15.987	18.307	20.483	23.209	25.188
11	2.603	3.053	3.816	4.575	5.578	17.275	19.675	21.920	24.725	26.757
12	3.074	3.571	4.404	5.226	6.304	18.549	21.026	23.337	26.217	28.299
13	3.565	4.107	5.009	5.892	7.042	19.812	22.362	24.736	27.688	29.819
14	4.075	4.660	5.629	6.571	7.790	21.064	23.685	26.119	29.141	31.319
15	4.601	5.229	6.262	7.261	8.547	22.307	24.996	27.488	30.578	32.801

From Donald B. Owen, *Handbook of Statistical Tables* (Reading, MA: Addison Wesley Longman, 1962). Reprinted with permission of the publisher.